CW01261341

THE HANDS-ON PROJECT OFFICE

Guaranteeing ROI and On-Time Delivery

OTHER AUERBACH PUBLICATIONS

The ABCs of IP Addressing
Gilbert Held
ISBN: 0-8493-1144-6

The ABCs of LDAP: How to Install, Run, and Administer LDAP Services
Reinhard Voglmaier
ISBN: 0-8493-1346-5

The ABCs of TCP/IP
Gilbert Held
ISBN: 0-8493-1463-1

Building an Information Security Awareness Program
Mark B. Desman
ISBN: 0-8493-0116-5

Building a Wireless Office
Gilbert Held
ISBN: 0-8493-1271-X

The Complete Book of Middleware
Judith Myerson
ISBN: 0-8493-1272-8

Computer Telephony Integration, 2nd Edition
William A. Yarberry, Jr.
ISBN: 0-8493-1438-0

Electronic Bill Presentment and Payment
Kornel Terplan
ISBN: 0-8493-1452-6

Information Security Architecture
Jan Killmeyer Tudor
ISBN: 0-8493-9988-2

Information Security Management Handbook, 4th Edition, Volume 1
Harold F. Tipton and Micki Krause, Editors
ISBN: 0-8493-9829-0

Information Security Management Handbook, 4th Edition, Volume 2
Harold F. Tipton and Micki Krause, Editors
ISBN: 0-8493-0800-3

Information Security Management Handbook, 4th Edition, Volume 3
Harold F. Tipton and Micki Krause, Editors
ISBN: 0-8493-1127-6

Information Security Management Handbook, 4th Edition, Volume 4
Harold F. Tipton and Micki Krause, Editors
ISBN: 0-8493-1518-2

Information Security Policies, Procedures, and Standards: Guidelines for Effective Information Security Management
Thomas R. Peltier
ISBN: 0-8493-1137-3

Information Security Risk Analysis
Thomas R. Peltier
ISBN: 0-8493-0880-1

Interpreting the CMMI: A Process Improvement Approach
Margaret Kulpa and Kurt Johnson
ISBN: 0-8493-1654-5

IS Management Handbook, 8th Edition
Carol V. Brown and Heikki Topi
ISBN: 0-8493-1595-6

Managing a Network Vulnerability Assessment
Thomas R. Peltier and Justin Peltier
ISBN: 0-8493-1270-1

A Practical Guide to Security Engineering and Information Assurance
Debra Herrmann
ISBN: 0-8493-1163-2

The Privacy Papers: Managing Technology and Consumers, Employee, and Legislative Action
Rebecca Herold
ISBN: 0-8493-1248-5

Securing and Controlling Cisco Routers
Peter T. Davis
ISBN: 0-8493-1290-6

Six Sigma Software Development
Christine B. Tayntor
ISBN: 0-8493-1193-4

Software Engineering Measurement
John Munson
ISBN: 0-8493-1502-6

A Technical Guide to IPSec Virtual Private Networks
James S. Tiller
ISBN: 0-8493-0876-3

Telecommunications Cost Management
Brian DiMarsico, Thomas Phelps IV, and William A. Yarberry, Jr.
ISBN: 0-8493-1101-2

AUERBACH PUBLICATIONS
www.auerbach-publications.com
To Order Call: 1-800-272-7737 • Fax: 1-800-374-3401
E-mail: orders@crcpress.com

THE HANDS-ON PROJECT OFFICE

Guaranteeing ROI and On-Time Delivery

RICHARD M. KESNER

AUERBACH PUBLICATIONS

A CRC Press Company
Boca Raton   London   New York   Washington, D.C.

Library of Congress Cataloging-in-Publication Data

Kesner, Richard M.
    The hands-on project office : guaranteeing ROI and on-time delivery / Richard M. Kesner.
       p. cm.
    Includes bibliographical references and index.
    ISBN 0-8493-1991-9 (alk. paper)
    1. Information technology. 2. Project management. I. Title.

HD30.2.K464 2003
004'.068'4—dc22   2003055771

This book contains information obtained from authentic and highly regarded sources. Reprinted material is quoted with permission, and sources are indicated. A wide variety of references are listed. Reasonable efforts have been made to publish reliable data and information, but the author and the publisher cannot assume responsibility for the validity of all materials or for the consequences of their use.

Neither this book nor any part may be reproduced or transmitted in any form or by any means, electronic or mechanical, including photocopying, microfilming, and recording, or by any information storage or retrieval system, without prior permission in writing from the publisher.

The consent of CRC Press LLC does not extend to copying for general distribution, for promotion, for creating new works, or for resale. Specific permission must be obtained in writing from CRC Press LLC for such copying.

Direct all inquiries to CRC Press LLC, 2000 N.W. Corporate Blvd., Boca Raton, Florida 33431.

Trademark Notice: Product or corporate names may be trademarks or registered trademarks, and are used only for identification and explanation, without intent to infringe.

Visit the CRC Press Web site at www.crcpress.com

© 2004 by CRC Press LLC
Auerbach is an imprint of CRC Press LLC

No claim to original U.S. Government works
International Standard Book Number 0-8493-1991-9
Library of Congress Card Number 2003055771
Printed in the United States of America 1 2 3 4 5 6 7 8 9 0
Printed on acid-free paper

DEDICATION

to Susie (-:

CONTENTS

Web Site Table of Contents xi
Preface xvii
Introduction xxiii

1 The Three Pillars of IT Delivery — Problem Resolution, Service Requests, and Projects 1
  Introduction 1
  The Business Context 2
  The Internal Economy for Investing in IT Services and Projects 3
  The Three Pillars of IT Delivery 7
  Managing Service Delivery 10
  Managing Project Commitments 16
  IT Metrics and Reporting Tools 20

2 The Project Management Office Business Model 25
  Introduction: Revisiting the IT Organization 25
  IT Service and Project Delivery Roles 33
  The Role of the Project Management Office: Measuring its ROI ... 42
  The PMO Value Proposition: An Initial ROI Estimate 50

3 Alignment and Planning — Doing the Right Things 57
  Introduction 57
  Getting the Business to Set IT Priorities 60
  Getting to "Yes" in Setting IT Priorities: An Approach to Business
  and IT Alignment 67
  Documenting and Accounting for IT Priorities:
  The Action Planning Process 73

4 Modeling and Managing Service Delivery 89
  Introduction 89
  Modeling Service Delivery Management 93
  The Service Delivery Agreement 100

vii

Basic SLA Terms and Definitions 101
Roles and Responsibilities of Process Participants...................................... 103
Representing IT Assets and Costs 105
Problem Resolution and Service Delivery Workflows................................. 107
Reporting on Results.. 108
Closing Comments 113

5 Project Delivery and the Project Management Life Cycle....115
Introduction 115
What Is an IT Project? What Is Project Management? Why Bother?........ 118
The IT Project Management Life Cycle — A Brief Overview 120
The Commitment Process 129
Project Delivery — Measurement and Reporting 139
The Role of the Project Management Office in Project
Management Services... 144

6 Collecting and Capturing Business Requirements
for IT Projects 147
Introduction 147
Preparing for Business Requirements Gathering... 150
Business Process Mapping 154
Process Decomposition 155
The Roles and Responsibilities Matrix 158
Process Rules.. 163
Performance Metrics 164
Process Templates and Tools... 166
Building a Final Picture of the IT Solution for the Customer................... 167
Closing Comments 168

7 Managing Lessons Learned — The Reuse and Repurposing
of IT Organizational Knowledge: A Case Study......................171
Introduction 171
The Whats, Whys, and Wherefores of KM.. 173
Getting Started: Introducing the Case Study 174
Business and Technical Requirements: Analysis and Design 178
The Development Process: Constructing Content
and Service Components.. 184
Certification, Launch, and Release... 193
Ongoing Operations 194
Lessons Learned 196

8 Architecting Success — The Role of Sensible IT
Architecture Management in Successful Service Delivery:
A Case Study...199
Introduction 199
Framing the IT Architecture Planning and Management Process............. 201
Introducing the Case Study 202

The Underlying Assumptions of an Enterprise's IT Strategy
and Architecture 203
The Process of Building and Maintaining an IT Architecture 207
Putting the Architecture Process to Work — IT Planning
and Procurement 214
Conclusions and Lessons Learned 220

9 Conclusions — The ROI of the PMO 225
Introduction 225
The ROI Discussion 228
Executive Support Services 231
Supporting Service Delivery 233
Supporting Project Delivery 236
Leveraging Technical Knowledge 238
Staff Support and IT Organization Culture 239
One Last Look at the PMO 241

Appendices

Key Templates from the PMO Tool Box

Appendix A: IT Project Justification Template 247

Appendix B: IT Annual Plan Template 253

Appendix C: PMO Value Calculation — Model and Template 263

Appendix D: Service Level Agreement Template 271

Appendix E: Project Management Life-Cycle Framework 283

**Appendix F: Project Leadership Questionnaire
for Change Management Projects** 295

Appendix G: IT Project Risk Management Matrix 301

Appendix H: Commitment Document Template 305

Appendix I: Master Project Schedule Template 317

Appendix J: Glossary 321

Appendix K: Selected Readings 329

Index 335

WEB SITE TABLE OF CONTENTS

The CRC Press Web site (http://www.crcpress.com/e_products/downloads/download.asp?cat_no=AU1991) complements *The Hands-On Project Office* by providing a complete electronic library of the tools, templates, and examples cited in the text. Its aim is to enable the reader to move swiftly and easily from an understanding of the book's content to the actual application of that knowledge in the workplace. The Web site is organized to align with the text. Each document label begins with the chapter where that tool is first mentioned. To this is added a descriptive tag. To show how these templates are to be employed, the author has included examples of completed templates, as well as the actual blank framework. Below, each Web site entry is listed, along with a brief description of its purpose. Feel free to adapt and use these tools as you see fit.

- **chpt0~1~value chain template** A PowerPoint slide of the Michael Porter value chain template for the mapping of the role of IT within the enterprise.
- **chpt0~2~enterprise transformation models** Both versions of the enterprise transformation diamond discussed in the Introduction and Chapter 3. The first considers the forces of change impacting the enterprise, and the second looks more specifically at the dynamics of change as they impact IT investment planning. The reader may need to adapt these models to reflect more accurately the particular realities of his or her business.
- **chpt1~1~internal economy model** The first slide offers a PowerPoint version of the resource stack discussed in Chapter 1. The second slide compares and contrasts service level agreement (SLA) expenditures

and project expenditures in the context of the internal IT economy model. These may be adapted to meet the reader's needs.

chpt2~1~IT Organization~model This attachment includes two simple IT organization charts: one without and one with a project management office (PMO) reporting to the CIO. The reader may adapt these illustrations to reflect the current or planned organization of his or her own IT team.

chpt2~2~IT Competencies~model 1 Originally developed for the IT organization of New England Financial (now part of MetLife), this model characterizes roles and responsibilities within IT in terms of a spectrum of operational to strategic competencies. This model will help the reader assess the available skill base of his or her organization and hence the gaps that might be addressed by the creation of a PMO.

chpt2~3~IT Competencies~model 2 Originally developed for the IT organization of Northeastern University, this model considers role-based competency requirements. This model will help the reader assess the available skill base of his or her organization and hence the gaps that might be addressed by the creation of a PMO.

chpt2~4~PM roles and responsibilities~example Another perspective on project management roles and responsibilities, as developed by Pat Erickson, manager of the Information Services (IS) Division's PMO at Northeastern University.

chpt2~5~PMO Operations~framework and model These two slides provide a functional picture of PMO operations and the roles and responsibilities of PMO personnel.

chpt2~6~PMO work alignment~model These slides model in greater detail the range of potential activities encompassed in PMO service delivery.

chpt2~7~pmo~job descriptions For the adaptation and use by the reader, templated job descriptions of possible project management roles within the PMO.

chpt2~8~km~job descriptions For the adaptation and use by the reader, templated job descriptions of possible knowledge management (KM) roles within the PMO.

chpt2~9~PMO value calculation~model and template This template models the PMO value proposition and allows for the calculation of PMO costs and benefits as they relate to the IT organization.

chpt3~1~IT~inventory~example As applied in a particular business setting, this tool captures information about the status of major IT project work and needs across the enterprise.

chpt3~2~IT~inventory~template A blank inventory template for the adaptation and use by the reader.

chpt3~3~priority worksheet~template A simple form for capturing IT project priorities once an agreement has been struck among IT's customers, internal service delivery teams, and external partner providers.

chpt3~4~IT Project Justification~template A more complex form for reflecting the total cost of ownership (TCO) for a given project, including its impact over several fiscal-year quarters. The form also includes a justification checklist to assist the user in preparing the supporting documentation for a proposed IT project.

chpt3~5~Action Plan~example A completed IT Action Plan document, including IT team updates and management review comments. This example illustrates how a fairly typical IT organization might employ the Action Plan tool. Note how the document focuses on the alignment between IT activities and enterprise goals and objectives.

chpt3~6~Action Plan~template A blank inventory template for the adaptation and use by the reader.

chpt3~7~planning~timetable~example A completed timetable for the various business processes associated with IT planning, such as budget cycles, internal/external business cycles, and various governance approval cycles. This master timetable helps harried IT executives keep track of multiple management process deadlines.

chpt3~8~planning~timetable~template A blank inventory template for the adaptation and use by the reader.

chpt3~9~annual planning process~workflow This slide serves as a graphic document of a planning process that may be adapted to reflect the particulars of the enterprise's processes.

chpt3~10~performance review~template Ultimately, the only way to ensure consistent observance of operating principles, best practices, and delivery management within the IT team is to embed these values into the IT organization's rewards and recognition system. This template incorporates these elements as part of a performance review form for IT personnel and may be adapted to suit the priorities of the reader's IT team.

chpt3~11~IT Board charter~example This attachment illustrates the type of document employed by the author to enroll members in an advisory board for an IT organization.

chpt3~12~Off-Cycle Approval Process~workflow This workflow addresses projects that enter the review process after the annual plan has been finalized by enterprise management.

chpt4~1~Service Delivery Workflows with Metrics~example This set of workflows for service delivery models the performance of help desk, field support, and technical teams and includes the measures that are tracked by the problem ticket system. This approach captures valuable performance data when the IT organization is rigorous in its use of its problem tracking system.

chpt4~2~ Service Level Management~example with templates As part of getting all of Northeastern University's IT service providers on the same page, IS adopted performance standards and metrics as a team process and then employed staff training to educate each and every member of the IT staff about how his or her work would be measured according to these new standards.

chpt4~3~SLA~template This is the template that the author employed both at MetLife and Northeastern University.

chpt4~4~SLA~example Here is how one organization employed the SLA template for its purposes.

chpt4~5~customer satisfaction measures~example This illustration outlines a series of metrics for typical IT services. Note that all metrics focus on measures of customer satisfaction.

chpt4~6~monthly service delivery report~template This template applies and presents the results of the aforementioned metrics in the context of a monthly internal management report for the IT organization. It is to be used for continuous self-improvement.

chpt4~7~monthly service delivery report~example This file demonstrates the use of the service delivery report in a real-life setting.

chpt5~1~project management life cycle~graphic Employ this graphic as you wish on all PMO documents as a reminder to your team and others of the steps in the project management process.

chpt5~2~project phases~model This detailed view of the author's project management framework includes references to roles, responsibilities, and project documentation and can be readily adapted to the reader's particular project management needs.

chpt5~3~project management process~work flows A graphic representation of the project management process as presented in the text.

chpt5~4~project management process~example with work flows The same project management phases model and workflow as applied to the project management process within Northeastern University's IS division.

chpt5~5~commitment~template This document complements the commitment articulation process. The full document includes a number of signoff pages that may be dropped from use by those organizations preferring a less formal project review and approval process.

chpt5~6~commitment~example This fleshed-out commitment document provides an example of how the tool works in practice.

chpt5~7~project leadership readiness~template This questionnaire serves as a simple tool to diagnose whether your project leadership has properly prepared the ground for the IT project at hand. Typically, this tool is most appropriate when the IT deliverables must be married with business process change for the desired result.

chpt5~8~risk management~template This tool allows the reader to rate and score the various risks associated with a particular IT project. When one is faced with options, this tool allows the reader to compare and contrast choices from the standpoint of risk management.

chpt5~9~facilities project delivery~template This tool includes a comprehensive checklist, for the management of facilities and infrastructure work associated with IT projects, and a spreadsheet for calculating related IT costs.

chpt5~10~Infrastructure Questionnaire for Projects~template This tool draws out the infrastructure-related implications of a system project so that hardware and networking issues may be brought to the attention of those responsible early in the project's life cycle.

chpt5~11~ Security Questionnaire for Projects~template This tool draws out the security-related implications of a system project so that hardware and networking issues may be brought to the attention of those responsible early in the project's life cycle.

chpt5~12~Project Issues and Action Items~template A template for recording and tracking risks and any other project-related issues.

chpt5~13~Project Issues and Action Items~example An example of an issue list in use.

chpt5~14~system launch checklist~template This tool, as originally developed by my colleague Pat Laughran, serves as a useful reminder of what must be in place as you launch a new information system or service.

chpt5~15~project scorecard~template A project scorecard template ready for adaptation and use.

chpt5~16~scorecard~examples Various project scorecards demonstrating how to reflect various project statuses.

chpt5~17~monthly project status report~template A template for the B or project side of the monthly operations review.

chpt5~18~monthly project status report~example An example of the template in use for the B or project side of the monthly operations review.

chpt5~19~master project schedule~template A template for the summary capture of all IT project work.

chpt5~20~master project schedule~example An example of how one might employ the project master schedule in a real IT business setting.

chpt5~21~project engineering framework~model A model for adaptation by the reader of the scoping of PMO project delivery activities within an IT shop.

chpt6~1~process map~template The author's process mapping tool template.

chpt6~2~process map~selling~example An example of how the process mapping tool may be applied to a sales process.

chpt6~3~process map~execution~example An example of how the process mapping tool may be applied to a service delivery process.

chpt7~1~pm meets km~model Initially developed by KPMG and then modified by the author to complement his project delivery and knowledge management approaches, this model illustrates the various levels of service and the connections between a knowledge site and a PMO.

chpt7~2~Web asset inventory~example An example of how one might map the functionality of a Web site as part of the analysis and design phases of the project.

chpt7~3~taxonomy~example This attachment includes three complementary pieces of analysis. The first includes an IT document inventory and the various relevant descriptors for each document type. The second defines document tag terms. The third provides a framework for the classification of documents within IT organizations workflows of service delivery, project delivery, and other business.

chpt7~4~Document Management~workflows This presentation documents the process for adding or revising content within the Northeastern University IS knowledge portal.

chpt8~1~architecture Web site~example This slideshow captures all key screens from the author's NEF/IS IT architecture Web site. These slides illustrate the design and layout of the various knowledge management components of this application.

chpt8~2~architecture audit~example As part of the architecture process, the author's team at NEF/IS regularly audited the progress of work by domain teams and the currency and completeness of their information in the IT architecture knowledge store. This slideshow captures the results of one such audit.

chpt9~1~the PMO Value Proposition~model This framework slide was presented in Chapter 5 as part of the project management discussion. Here, the model frames the discussion of the PMO's return on investment (ROI).

PREFACE

There was a time not so long ago when information technology (IT) executives labored in obscurity and the IT organization was viewed by its enterprise sponsors as a necessary but expensive piece of overhead. The best run IT shops kept costs down and the presses rolling. The broad spectrum of IT's strategic alignment with the business was not of concern, and only rarely did one find a chief information officer (CIO) sitting on the executive management committee. Today, the situation is quite different. Whereas in the early 1990s IT executives may have fretted about having a role in the enterprise's strategic visioning, today IT enablement serves as a core component in most business plans. As a result, proportionate spending on IT has grown significantly, and the CIO is now viewed as a key contributor to the enterprise's business strategy and planning processes.

For IT professionals, these developments are a double-edged sword. On the one hand, IT staffing and resource levels have increased, as has the status of the IT organization within the greater enterprise. Expectations run high that IT products and services will give the business a competitive edge, improve productivity, allow for greater complexity and diversity in business operations, and lower overall operating costs. On the other hand, the pressure is on IT to deliver. The business side of the house now anticipates positive, significant outcomes from technology spending. Indeed, business managers expect measurable returns on their growing investment in IT, even as they request and fund grand, ever more complex and risk-laden IT projects. As the past decade of huge IT investments has sadly demonstrated, not all of these expectations have been well informed, leaving many chief executives dissatisfied with their IT organizations and the substantial enterprise resources these now consume.

The primary premise of this book is that the greatest challenge faced by IT executives today is in demonstrating value-added contributions to the enterprise of their technology organizations. To be specific, IT management is now pressed more than ever to deliver a quantifiable return on the investments (ROI) made in corporate IT. Sadly, many IT projects continue to come in late and over budget, while falling short of customer expectations. For that matter, the bread and butter of day-to-day IT service delivery is all too often left to its own devices without a serious consideration of how process improvements could lead to performance improvements and greater customer satisfaction. The primary value of *The Hands-On Project Office: Guaranteeing ROI and On-Time Delivery* is to provide carefully considered and thoroughly tested processes, techniques, and tools that the harried IT manager may apply immediately to improve the delivery of IT products and services. In brief, this book serves as a compendium of best practices and practical recommendations for the consideration and use by IT service delivery and project management professionals.

Now, it is true that the existing trade literature is populated with books to assist IT practitioners in the execution of their day-to-day responsibilities. Most of this body of work, however, considers the minutiae of project management, neglecting entirely the service side of the equation, or focuses on the grander issues of business/IT strategic alignment. As good as many of these books are, by and large they do not address the particular needs outlined. In the first place, their continuing emphasis on the debate over whether IT should align with the enterprise is no longer necessary. Almost anyone in an organizational leadership role would agree on the importance of such an alignment within the enterprise. To that point, many business executives are now sufficiently conversant in the whys and wherefores of information technology to set the IT direction for their own enterprises, leaving the CIO to implement their vision. Second, many IT management texts are entirely too academic, providing complex and cumbersome models and methodologies that would never be considered — let alone employed — by experienced, overworked practitioners in the field. Appendix K provides a list of recommended readings of the best and most current books in these areas.

From the outset, this book takes a different tack. It serves as a practitioner's handbook, providing simple, deployable frameworks, practical tools, and proven best practices for successful IT service and project delivery management. All of the tools mentioned in the text and captured in the accompanying Web site (http://www.crcpress.com/e_products/downloads/download.asp?cat_no=AU1991) have been field tested in organizations ranging from colleges and universities to major for-profit corporations to small, entrepreneurial enterprises and so-called dot.coms by the author who, for twenty years, has served as a CIO, a chief knowledge officer (CKO), or an IT consultant. Although each tool or approach is described in the body of the text as it is to be applied in the field, the author has also provided by way of illustration a companion document library on the accompanying Web site, which includes both actual tool templates and examples of completed forms. Together, these examples illustrate how the practical advice of *The Hands-On Project Office* may be adapted and applied directly to the reader's own management situations.

What enterprise executives really want from their IT organizations is a reasonable level of economy, consistency, and reliability in the execution of IT assignments. Most business leaders already recognize the paramount role that IT plays in serving the business side of the house and its customers. What they do not see is a satisfactory return on their investments in technology-enabled business processes, and in some instances, even positive outcomes for the funds expended. IT consulting firms and trade magazines are now all singing from the same hymnal that ROI is the name of the game for corporate IT expenditures in 2003 and beyond. Indeed, in the current, tough economic climate the mantra is to do more with less. This measure is now applied to the IT organization, just as it is has been employed for some time against the performance of other key business units within the enterprise. In today's competitive marketplace, cost cutting must go hand-in-hand with higher output and greater productivity. This means that IT organization performance will come under ever greater scrutiny.

IT leaders must get proactive about doing more with less while demonstrating greater success in addressing customer requirements. *The Hands-On Project Office* shows IT managers how to better coordinate their work efforts, hold their own teams accountable, communicate the impact of IT overall product and service delivery to the enterprise, and more generally, demonstrate the value of the IT organization to the whole. The challenge in these scenarios is to establish the means for repeatable IT team successes. Throughout, the book's approach is to address these very objectives by streamlining the work processes that capture and reflect information concerning IT delivery, and that clarify associated roles, responsibilities, customer expectations, and performance measures. In short, this book provides immediately implementable solutions to IT personnel faced with the day-to-day management of large, complex assignments. The audience for this book includes IT executives and line managers, customer service and call center personnel, project managers and business analysts, knowledge management (KM) officers, IT architects, and other senior technical and business managers across the enterprise.

In preparing this text, the author has enjoyed and benefited from interactions with a large number of private and public sector IT managers. To acknowledge them individually would require a book in itself. The author therefore thanks them and their respective organizations collectively for their time and the quality of their input into his research. They include Arkwright Insurance, AT&T, Babson College, the Bank of Montreal, the Bank of Nova Scotia, Bankers Trust (New York), BankVest Capital, Boston University, Bryn Mawr College, Camp Dresser & McKee, Columbia University, Deloitte & Touche, Disney Enterprises, Drexel University, Earley and Associates, The Faxon Company, Harvard University Medical School, Hofstra University, The Hurwitz Group, ITechnology, KPMG Consulting Services, Marriott Corporation, MIT, MetLife, Multibank Financial Corporation, The New England, Northeastern University, Paine Webber International, The Robbins Company, SilverPlatter Information, Simmons College, the State of California's Secretary of State's Office, the Treasury Division (Canadian Federal Government), University of Ottawa, University of Pennsylvania, Wheelock College, and William & Mary University.

The author also acknowledges the assistance of a number of consulting firms with established expertise in change management, namely Axion Consulting, Beacon Application Services, Boston Consulting Group, CSC Index, and Symmetrix, who have willingly shared their expertise with me. In particular, I am grateful to the folks at Symmetrix for introducing me to a strategy for the effective marketing of change within the organization, and to Beacon Application Services for demonstrating the use of new groupware tools in the rapid prototyping of new IT-enabled business processes.

As helpful as these people have been in shaping my views on managing IT services and projects, nothing has proved more formative than my own field experience. The list of enterprises with whom I have been associated as an IT executive, a business strategist, or a technology planning consultant include AIRS Software, Arkwright Mutual Insurance, Babson College, the California State Archives, Camp, Dresser & McKee Consulting Engineers, East Tennessee State University, The Faxon Company, Harvard University Medical School, Kingsport City Municipal Government, the Merrill Palmer Institute, MetLife, Multibank Financial Corporation, New England Financial Corporation, Riemer & Braunstein Attorneys at Law, The Robbins Company, SilverPlatter Information, Simmons College, Trellis Computer Services, Wayne State University, the WGBH Foundation, and Wheelock College. This collection of organizations encompasses large and small, public and private, and high-tech and low-tech enterprises. But they all have two things in common: a need to manage their respective IT services more effectively and efficiently, and a need to invest in and restructure their operations and uses of IT to better address the evolving requirements and expectations of their paying customers. As I have worked for this diverse group of organizations over the years, I have come to appreciate both the common and the unique strategies and techniques required to achieve successful enterprisewide IT service and project delivery. For these close and rich affiliations, I am most grateful. Without them, I would not have been in the position to prepare this book.

Lastly, I thank my many colleagues at New England Financial (NEF) and Northeastern University (NEU) for their support and direction in laying the foundation for this manuscript. In particular, I recognize the NEF project office and E-commerce development teams that I led, the NEU information services management team led by Bob Weir and Rick Mickool, the NEU Project Management Office team led by Patricia Erickson, and the NEU Quality Assurance and Release Management team led by Scott Wiersman for their efforts in testing my models, validating my ideas, and correcting my errors. I also recognize Beth Anne Dancause, Pam Marascia, Denise Siciliano, Glenn Hill, and Steve Theall for their contributions to my management frameworks, process modeling, and tool development. Thank you all so very much!

To all of these good people, I must add the names of three who through their good humor, affection, and love have sustained me during the trying process of turning my ideas into this book, namely my sons Henry Alexander and Samuel Benjamin Kesner, and my wonderful, supportive, and ever patient spouse, Susan Nayer Kesner. Your presence has kept me on track and on schedule. This book is a tribute to my colleagues' and my family's collaborative efforts. My readers have them to thank for what is of worth in the pages that follow, and the author alone to blame for any errors in fact or judgment found herein.

Richard M. Kesner
Northeastern University

INTRODUCTION

> To exist is to change, to change is to mature, to mature is to go on creating oneself endlessly.
>
> — **Henri Bergson**
> *French philosopher (1859–1941)*

Today's organizations are beset by the forces of transformation. The scope and pace of changes are staggering. Markets and customer requirements appear to redefine themselves on an almost daily basis. Global politics and domestic legislation continue to rewrite the rules on how to play the game — creating new opportunities as they shut down established practices. Mergers and acquisitions, the rise or fall of competitors, and the emergence of new players all have contributed to a feeling of constant churning in the business environment. Even government institutions and other not-for-profit organizations are feeling the heat as they are closed down, significantly restructured, or privatized. Through the introduction, seemingly overnight, of new information technologies, established business processes and entire enterprises are radically altered or driven out of existence.

In this sea of change, excellence in the deployment of information technology (IT) has emerged as a strategic necessity. Those enterprises that demonstrate a high level of competence in leveraging IT to serve their customers, collaborate with their upstream suppliers and their downstream distributors, and enable their own employees also tend to excel as performers in their respective industries. How critical is the role of IT in your particular business? To answer this question, Michael Porter's value

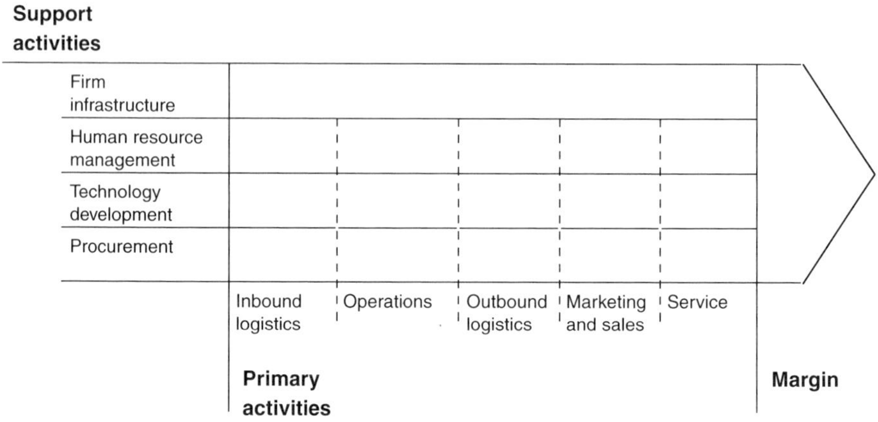

Support activities

Firm infrastructure					
Human resource management					
Technology development					
Procurement					
	Inbound logistics	Operations	Outbound logistics	Marketing and sales	Service

Primary activities

Margin

Exhibit 1 Michael Porter's Value Chain Model for Business Process Analysis

chain model helps to assess the ways in which IT products and services enable your enterprise.* See Exhibit 1.**

The Porter model emphasizes the value chain of business processes whereby each business activity has a so-called upstream and downstream partner in delivering value to the customer. Porter employs a manufacturing metaphor for his process model, using such terms as inbound logistics, operations, outbound logistics, marketing and sales, and service to define process steps. But the reader may easily extrapolate in applying this framework to his or her own organization. Porter's model will help you distill the essence of your enterprise's key business processes, how they interconnect, and where IT plays or should play an enabling role.

The Porter framework readily applies to the IT organization itself, where each and every service or project deliverable requires the collaboration and coordination of numerous IT operating units, partner providers (external and internal to the organization), and more often than not customers and other business process stakeholders. To this chain of events, Porter's model adds a framework for the consideration of support activities that are internal to the enterprise and enable delivery:

---

* The source of this analytical tool is still relevant and a classic of business analysis. See Michael Porter and Victor E. Mylar, "How Information Gives You Competitive Advantage," *Harvard Business Review* 63, no. 4 (1985): 149–160.
** For an electronic version of the value chain template, see *The Hands-On Project Office*, http://www.crcpress.com/e_products/downloads/download.asp?cat_no=AU1991, chpt0~1~value chain~template.

- Firm infrastructure — management overhead, real estate, financing, and so forth
- Human resource management — the people who do the work
- Technology development — information resource and technology management, but also product/service research and development
- Procurement — obtaining the raw materials for production

Again, the metaphor of the value chain is extremely useful in that it emphasizes the linkages between the individual activities of the business process. A chain is not complete, nor does it serve a constructive purpose, unless all its links hold. Similarly, a business process must be viewed in its entirety and not as a set of discrete functions. The continuous flow from inbound logistics to post-sales servicing and the underlying support structures, in concert, delivers value to the customer. We need to view our own enterprises in this comprehensive manner and ask ourselves where and how does IT best contribute to overall customer value in each intersecting box of Porter's value chain matrix.

Indeed, Porter observes that successful enterprises add value at each link along the value chain and that business operations can only be appreciated in terms of the totality of the processes that they encompass. Thus, if a function is not adding value to the process, it should be eliminated. If a function's value lies in facilitating corporate performance downstream (i.e., later in the process), its contributions must be evaluated in terms of their support of downstream deliverables. Given Porter's simple yet comprehensive view of business processes, it has received wide acceptance among academics and managers alike. For our purposes, Porter's model facilitates an enterprisewide assessment of IT impacts and opportunities, as well as the components of the value proposition for measuring IT organization performance.

However, the true potential IT enablement within your organization may depend less on its value chain roles and more on the business contexts and culture within which your enterprise operates. First, although developments in IT can influence an enterprise to transform itself, IT does not bring about change. People do. For that matter, organizations can improve themselves incrementally or seize upon more radical solutions without recourse to IT. Second, the deployment of technology is no assurance that an enterprise will become more effective or efficient, far from it. When IT is employed for its own sake and business processes are merely automated without significant adaptation, the result is more often than not a faster, more expensive, and still broken business process. Third, many organizations have IT infrastructures employing mainframe-based systems and so-called enterprise resource planning (ERP) software suites. These pose a significant barrier to change because they cost so much to install initially

— and an even greater amount if they are to be enhanced over time. With these legacy systems in place, it sometimes seems that one must move mountains to achieve modest performance improvements.

Thus, if IT truly is to enable the overall performance of the enterprise, the purchase and introduction of these technologies must integrate with redesigned business processes. Newly added technologies must be flexible, adaptable, and easy to use. Unfortunately, repeatable success in the merger of IT and business processes is hard to come by. For most technologists, this means getting more involved in the less technical side of the business, drawing upon competencies that may not be part of the typical stock in trade of IT teams. As you will see in the subsequent chapters of this book, the ongoing challenge is not to select the right information technologies, because this fit will vary with each business setting and the unique characteristics of the context and culture of your enterprise. Instead, *The Hands-On Project Office* strives to reinforce the need to properly adapt and integrate IT service and project delivery disciplines into the preferred modes of operation within the larger, parent organization.

To help the reader better understand this broader framework and the highly dynamic environment within which the modern IT organization must position itself, a colleague at Babson College, Charlie Osborne and I have revised Harvard Business School's enterprise transformation diamond to reflect this view.[*] It shows the external environmental market forces alluded to previously (including the voice of the customer), as a backdrop to the internal dynamics of organizational change and IT investment decision-making. Within this context, the model represents the various factors at play in shaping the enterprise's business processes and hence its choices among enabling IT strategies. See Exhibit 2.[**]

As our model depicts, at the core of the enterprise are its business processes that drive the organization's IT direction and investment strategies. Because of the external environment's continuous influence, these business processes are never truly static. Customer demands, competitive pressures, and emerging technology innovations all play a part in forcing the enterprise and its IT capabilities to adapt and transform.

---

[*] Charlie Osborne served as a distinguished professor of Management Information Science at Babson College until his tragic and untimely death in 2000. He and I collaborated all too briefly on a training program for IT executives, offered through Babson's School of Executive Education. This analysis tool emerged from and was refined as part of that course development effort.

[**] For an electronic version of the transformation diamond, see *The Hands-On Project Office*, http://www.crcpress.com/e_products/downloads/download.asp?cat_no=AU1991, chpt0~2~enterprise transformation models.

Introduction ■ xxvii

```
                    ┌─────────────────┐
                    │   Management    │
                    │   procedures    │
                    │ • Management control
                    │ • Incentives    │
                    └─────────────────┘
┌──────────┐         ┌─────────┐         ┌──────────────┐
│Job design│  ◄───►  │Business │  ◄───►  │    Norms     │
│• Structure│         │ Process │         │• Informal rules
│• Sequencing│        └─────────┘         │• Organizational habits
└──────────┘                              └──────────────┘
                    ┌─────────────────┐
                    │  Information    │
                    │    systems      │
                    │ • Reporting/analysis
                    │ • Information distrib.
                    └─────────────────┘
```

The external environment - a.k.a. market forces

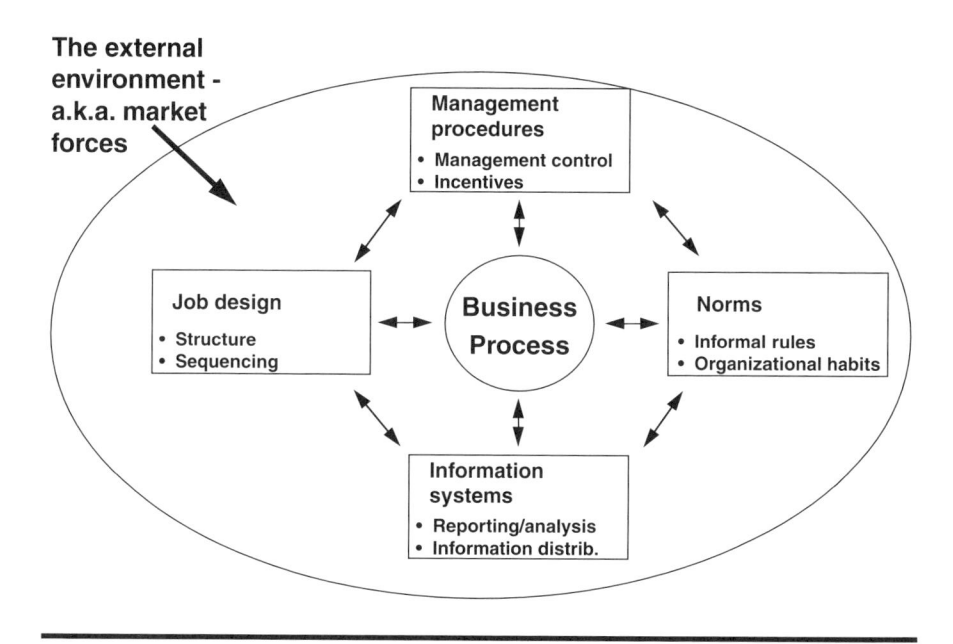

Exhibit 2   The Osborne/Kesner Enterprise Transformation Forces Diamond

Furthermore, as represented in the change diamond, internal forces are also at play. In the top quadrant of the illustration is a box for management procedures. This grouping of factors includes the organizational constructs of the enterprise, including chains of command, formal policies and procedures, salary structures and incentives, and so forth. The left quadrant encompasses job design and therefore such elements as the structure and definition of individual and team responsibilities, performance standards and measurement, the sequencing and relative importance of tasks, and the grouping of activities into functions and subprocesses. The right quadrant, labeled norms, refers to the enterprise's corporate culture. This grouping of factors includes all of the behavior patterns and informal rules condoned by the corporate community. In a more general sense, the norms quadrant of the model subsumes the work practices and styles of the enterprise. Finally, the information systems quadrant includes the IT capabilities and the IT-based practices of the organization, as well as its modes of information sharing and distribution.

The transformation model may be thought of as a mobile in which the movement of any quadrant influences all of the others. Thus, business process changes will influence management procedures, job design, norms, and in all likelihood the structure and operation of information systems. Changes in management procedures will affect norms, jobs design, and business processes; norms changes will influence management procedures, information systems, and business processes; and so forth. The entire model is highly dynamic, especially when one takes into account the environmental factors that influence the positioning of the overall enterprise.

Thus, the transformation diamond is an excellent representation of the factors and forces at play within the enterprise and between the enterprise and its external marketplace. This simple model should help to strengthen understanding of the dynamics of change and the need to anticipate its ripple effects as we move from the current to some future state — either proactively, through our own continuous improvement efforts, or reactively, as the environment around us forces change upon the enterprise. Similarly, no major investment in IT will prosper without due consideration of how the adoption of these technologies and the associated, perhaps radical, process changes will impact the various quadrants of this representational model. Fortunately, it is just these issues that the methodologies presented in this book will address. As we proceed, I suggest that readers employ tools, like the Porter value chain matrix and Osborne/Kesner change diamond model, to establish a broad frame of reference in evaluating the role that IT should play within their organizations. Be sure to refer to these frameworks as you assess the challenges in implementing the many suggestions and recommendations that follow.

As an enterprise IT manager, you already face the challenge of how best to direct the resources of your team to enable the success of the business. You may already have access to the best people and technologies, but are you satisfied with the quality of your team's delivery? *The Hands-On Project Office* will assist you and your team in delivering the greatest value to your corporate sponsors. To that end, it provides simple, field-tested business models, frameworks, and tools, designed expressly for IT organizations like yours.

What is critical to the success of your IT team and how can *The Hands-On Project Office* make a difference? The author would summarize an IT team's critical success factors as follows:

- Communication
  - Clear, continuous communication within the IT team and between that team and its customers and external partners, focused on clarifying IT roles, responsibilities, and deliverables
- Delivery management
  - Predictable delivery of products and services
  - Project management, subcontractor management, requirements management, and IT architecture management that leads to timely and cost-efficient implementation or operation of IT products and services

- Resource management
  - Resource flexibility deployed and allocated to highest-value customer business requirements
  - Ability to attract and retain top-performing staff who are aligned with enterprise business priorities
- Architected and managed solutions
  - Products and services, whether developed in-house or provided from external sources, that are well integrated with customer needs and the existing base of information technologies
  - Adherence to the enterprise's IT architecture and engineering standards and the consistent use of quality assurance (QA) processes
  - Baselining and ongoing measurement of performance and regular reporting to the appropriate stakeholders
- Collaboration
  - Partnering with the business owners of systems to ensure effective and efficient use of those systems and of overall system integrity and data quality
  - Partnering across IT to ensure optimal system and data integration
  - Partnering across the enterprise to ensure knowledge sharing around best practices and lessons learned

The Hands-On Project Office provides assistance in all of these areas. The book demonstrates how IT leaders might describe more effectively the value of their products and services to the enterprise at large. It recommends best practices in the areas of IT planning, priority setting, and service and project management. The book goes on to illustrate how IT personnel might establish a knowledge management (KM) platform to better communicate and leverage the technical, process, and people know-how from across the IT organization to improve ongoing team performance. Each major set of recommendations is accompanied by simple tools to jump-start IT efforts.

Chapter 1 creates an internal-economy model for IT organizations, systematically reviewing the operational, organizational, financial, and human resource dimensions of the typical enterprise-level IT operating unit. With this business model as context, Chapter 1 explores in more detail the various components of IT delivery and how best to model and measure performance excellence.

With Chapter 1 as a foundation, Chapter 2 formally introduces the concept of the project management office (PMO). This chapter discusses the roles and responsibilities within typical IT service and project delivery processes, and then considers the advantages of creating a center of excellence within the IT organization to oversee this work. Due consideration is given to readers' organizational contexts: larger IT organizations will most likely benefit from the presence of an actual PMO, while smaller IT shops will address this need through the training and development of line service and project delivery management personnel. Whether staffed or virtual, the PMO serves as the fulcrum for leveraging best practices within the IT organization. The balance of *The Hands-On Project Office* examines in closer detail particular processes that could operate out of the PMO for the benefit of the IT team.

The cognate of "doing things right" is "doing the right things." Therefore, Chapter 3 outlines a practical, hands-on alignment and planning process to ensure that IT efforts and resources are focused on work of the highest priority to the enterprise. This chapter includes tools for the creation and communication of enterprisewide IT planning and project and resource prioritization. It also introduces simple models for the effective communication of IT resource commitments to enterprise management. Throughout, the reader will find that the role of the PMO is to support the efforts of the IT management team and to record and refine the documents that emerge from their deliberations with their business colleagues.

Chapter 4 examines service delivery management best practices, including the design and implementation of service level agreements (SLAs), the measurement of service level performance, the maintenance of reporting processes, and more general support of the IT customer relationship management. PMO personnel provide the underlying support for these processes and maintain the templates, tools, metrics, and reporting mechanisms that are the hallmarks of effective service delivery management. This chapter also details the workings of the service management templates to be found in the PMO tool box in the appendices and on the complementary Web site.

Chapter 5 takes a similar approach for project management best practices, encompassing project scoping and commitment making, risk and resource management, the oversight of day-to-day project engineering and delivery processes, delivery measurement and reporting processes, and the more general coordination of IT project activity at the portfolio and enterprise levels. The PMO is essential to the success of project delivery more than to any other subject covered in this book. Chapter 5 provides any number of ways that the PMO might contribute to these challenging IT assignments. Throughout, the chapter includes a review of appropriate project management templates and tools to be found in the PMO tool box and on the Web site.

Of the two primary disciplines associated with PMOs, project management is perhaps the more obvious. However, business analysis, as associated with both project requirements gathering and process design and reengineering, is equally important. Indeed, these efforts are essential to understanding customer needs and managing customer expectations, as well as to implementing successfully a business-enabling IT solution. Chapter 6 offers a comprehensive tool set for discovering and documenting customer needs in ways that are comprehensible to nontechnical users but highly relevant for those assigned to build, operate, and service IT systems. Clearly, this is a PMO role. The challenge to those involved is how best to gather requirements without seriously impacting the ongoing obligations of your business partners.

In Chapter 7 and Chapter 8, the author draws upon his experiences as a chief knowledge officer (CKO) to illustrate how the tools of KM may be brought to bear in improving IT team performance. Indeed, to foster a truly successful IT organization, managers must find ways to harness and leverage the knowledge that resides within their own technical teams. Most of this knowledge is tacit, but some of it is documented in the artifacts (e.g., the commitment documents, the project plans, the business and technical requirements, the performance data, and so forth) that emerge from the activities of IT personnel. For that matter, in large enterprises, it is even difficult to discover where the expertise for particular IT systems may reside. Chapter 7 demonstrates how an IT organization can get its hands around its own communities of best practice. Here again, the author suggests that PMO personnel are best situated, as the creators or curators of the organization's process documentation, to provide this valuable KM function, promoting the reuse and repurposing of team know-how. This chapter also provides a simple design for an IT organization's KM Web site.

Chapter 8 builds on Chapter 7, demonstrating how the principles of KM may be brought to bear to manage information better concerning the flow of new technologies and IT assets within the enterprise. Although the responsibility for IT asset management cuts across the IT organization, the overall design and maintenance of the knowledge around the enterprise's IT architecture might best fall to the PMO as a complement to its other facilitation and KM responsibilities. For that matter, in their project delivery support roles, PMO personnel will be among the first to learn of changes in the organization's technical direction, and because of their involvement with IT's entire technology portfolio, they will have a better sense of the big picture. Chapter 8 provides practical advice on how best to leverage the services of the PMO to oversee IT investment strategies.

Chapter 9 reviews the primary rationale for creating the PMO and cites practical examples of how the investment in these support functions will benefit the enterprise. This final chapter sums up the findings of the book and its approach in assisting IT organizations achieve repeatable success in their service and project delivery efforts.

Finally, the appendices include hardcopy versions of the major tools discussed in detail throughout the book, while the accompanying Web site (http://www.crcpress.com/e_products/downloads/download.asp?cat_no=AU1991) holds an expanded set of models, templates, tools, and forms, as well as examples of the tools as employed in actual business settings.

As business and IT managers, readers should by now have a sense of what my book can do for them. From the outset, this work encourages IT leaders to focus on delivery to the customer. The book then provides IT practitioners and their business-unit counterparts with the partnering tools to identify and deliver work of the highest value to the enterprise and its customers. *The Hands-On Project Office* speaks to the need for metrics, accountability, and continuous communication. To that end, the book defines IT service and project delivery roles and responsibilities, management and measurement tools, reporting formats, and a host of other practical applications that help get the IT job done.

To successfully deploy the processes and tools described herein, however, the reader must hearken back to the analytical models presented at the opening of this Introduction, namely the Porter value chain matrix and the Osborne/Kesner enterprise transformation model. Please bear in mind that no two organizations are exactly the same. Circumstances range widely from enterprise to enterprise, from industry to industry, from location to location, and so forth. Only with a thorough understanding of one's own business context, can one determine what may work best or is even relevant to the situation. The reader is encouraged to maintain his or her objectivity while plowing through this volume. Where you find relevance, please do not take the author literally. Adapt my recommendations to the unique needs of your situation. Open your eyes to the possibilities, but also to the pitfalls, of applying my best practices in addressing the particular challenges of your own situation. Good luck!

1

THE THREE PILLARS OF IT DELIVERY — PROBLEM RESOLUTION, SERVICE REQUESTS, AND PROJECTS

INTRODUCTION

Although some information technology organizations operate as separate business entities, running on their own profit-and-loss statements, and although a small percentage of companies have outsourced their information technology (IT) operations to third-party service providers, the vast majority of today's enterprises maintain their own, internal information technology shops. In a world where IT plays an enabling role in almost every aspect of business activity, it should come as no surprise that both the investments and the expectations for the positive impact of information services within these enterprises are on the rise. For IT organizations, the challenge, as always, is how to satisfy the customer, given real-world resource constraints, the actual limitations of available information technologies, and most importantly, the business process and organizational barriers to change and innovation posed by the enterprise itself.

For the IT organization, the true measure of success is easily stated: the consistent and economical delivery of day-to-day services and project work. But how is this to be achieved? The purpose of this chapter is to provide a starting point for the discovery and maturation of an IT organization that excels at delivery. To begin, the author frames a practical model for depicting the so-called internal economy of the IT organization. The chapter then proceeds through a systematic consideration of how an information technology team should define, manage, and measure its deliverables in terms that are meaningful to the enterprise. Throughout, the approach focuses as much on a rigorous application of effective communication and shaping customer expectations as it does on the actual delivery of positive results. In the end, customer satisfaction may be realized, at least in part, through the use of the practices detailed in the text, combined with the associated and necessary internal and external partnering to deliver on time and within budget. A more detailed exploration of these methods and tools are found in subsequent chapters of this book.

THE BUSINESS CONTEXT

Today, every type of enterprise requires some level of information technology enablement. Indeed, one would be hard pressed to identify a corporate board or a chief executive who does not appreciate this need. For its part, the IT vendor community builds on this growing appreciation for the role of IT by overselling the value of investing while understating the total cost of ownership and the difficulties in bringing enabling technologies to bear. In truth, successfully integrating IT into business processes is a complicated, resource-intensive activity, requiring expert technicians working hand-in-hand with the business owners of those processes. Unfortunately, most enterprise business leaders do not want direct involvement in IT service or project delivery, and most information services professionals are not business process experts. Thus, from the outset, IT undertakings tend to be misaligned. Too much is promised; too much is assumed; and too much is left unsaid. These circumstances inevitably lead to misunderstandings, to scrap and rework, to poor results, and, hence, ultimately to unmet customer expectations.

In addressing these issues, it should be clear to the reader that the capacity of the underlying technologies is not the major problem. Most enterprises simply lack a comprehensive process that ensures the synchronization of IT service and project investments with overall business planning and delivery. Indeed, many enterprises fail to clarify and prioritize their information technology investments based on a hierarchy of business needs and values. Some enterprises do not insist that their IT projects have a line-of-business sponsor who takes responsibility for a project's outcome and who ensures sufficient line-of-business involvement in project delivery. Few, if any, are willing to commit to the internal changes within their own business processes, even though it is only through these transformations that they will realize the return on investment (ROI) on their IT investments. To address these shortcomings, each and every enterprise should embrace a process in which the business side of the house drives IT investment, in which both business and IT management holistically view and oversee IT project deliverables and service delivery standards, and in which ownership and responsibility for IT projects and services are jointly shared by the business and the IT leaderships.

For their part, IT organizations must adopt practices that focus on listening to the voice of the customer, confirming customer requirements, holding IT delivery teams accountable for their commitments, and improving overall communication within IT and between IT and its business partners. Furthermore, IT leaders must concern themselves with measuring and reporting on both the process of delivery and the team's actual deliverables. To these ends, it is essential that the IT organization clearly and explicitly define and manage projects and services so that both the sponsors and customers external to IT and the IT team itself are clear about what is to be done; by whom; when; at what cost; and how the ultimate measures of successful delivery are defined, understood, and employed by all stakeholders. Embracing these practices may run contrary to corporate culture, but they are nevertheless essential for the continuous improvement of the IT organization and for achieving consistency in serving its customers.

As a starting point, the author offers a simple model of the internal economy for information technology services that in turn drives IT's allocation of resources between service delivery and project work. For the IT executive who must manage the confluence of activities, the discussion boils down to how best to address the most impactful information technology needs of the enterprise within the constraints of the IT team's available resources. This internal economy model distills all of the complexity around resource allocations, so that management may more easily consider the larger issues of priorities and alignment with the business. This model also serves as the foundation for a more detailed consideration of two complementary IT business processes: service delivery management (Chapter 4) and project commitment management (Chapter 5). For the moment, let us delve into these activities in a little more detail. The author will then briefly introduce a series of high-level models for the more effective synchronization, communication, and oversight of IT service and project commitments, including the use of measurement and reporting tools. This précis will lay the groundwork for the rationale in creating a project management office (PMO), which will follow in Chapter 2.

THE INTERNAL ECONOMY FOR INVESTING IN IT SERVICES AND PROJECTS

All organizations are resource constrained. Therefore, their leaders must choose where best to invest these limited resources. With the growing use of information technologies across the enterprise, IT's share of the pie may be increasing. Nevertheless, this too has its limits, requiring planning and prioritization in line with the needs of the greater enterprise. To effectively and efficiently manage IT resources, one must understand the full scope of the demands driving the prioritization of these IT investments.

At the most fundamental level, organizations invest in technology in compliance with mandated legal and accounting requirements, such as those set forth by federal and state taxation authorities, government legislation, regulatory statutes, and the like. At the next level, an enterprise expends resources to maintain its existing base of information technology assets, including hardware and software maintenance; system licenses and upgrades; security services; and desktop, storage, and printer expansions and replacements. These investments are meant to "keep the lights on" and the business going, and therefore are not discretionary. Neither are these costs stagnant. They go up with inflation and as new workers are added or as the network and related IT infrastructures grow alongside the enterprise. Furthermore, as new IT services are introduced, they become, over time, part of the enterprise's embedded base of IT, expanding the nondiscretionary spending on technology.

Because none of these information technology products and services runs on its own or functions flawlessly, the IT organization must also provide significant and costly end-user, operations, and production support and troubleshooting. Similarly, because neither the requirements of the IT customer nor the evolution of information technology products themselves is static, there is a constant need to enhance existing IT products and services and to invest strategically in new IT capabilities. Thus, day-to-day delivery must balance ongoing services — typically running twenty-four hours a day, seven days a week (24/7) — with a wide range of service and system enhancements and new project work. Often, IT project delivery resources overlap with those focused on service delivery for the simple reason that the development team must understand the current state of the enterprise's business requirements and IT capabilities if it is to deliver requested improvements. Furthermore, it is a good practice to ensure that those maintaining an IT service have a hand in its creation or, at the very least, thoroughly understand its IT underpinnings. Thus, a balanced IT organization requires a work force dedicated to 24/7 service delivery and ongoing infrastructure maintenance, overlapping a core group focused on technological innovation, development, and systems migrations and integrations. In smaller IT shops, these teams will comprise the same people; in larger organizations, these teams may coexist as separate entities and may even compete against one another for scarce resources.

Taken together, these various layers of IT investment establish the boundaries of the IT organization's internal economy or what I like to refer to as the total cost of IT ownership. My model groups information technology expenditures into two large buckets: nondiscretionary costs that support existing IT investments and discretionary costs that fund new initiatives, including major system enhancements and new IT projects. Often, IT organizations will establish service level agreements (SLAs) to codify the annualized terms, conditions, and resource allocations associated with these nondiscretionary services. (See Chapter 4 for a detailed discussion of this process.) By contrast, discretionary costs may be treated on a discrete one-time basis and are typically governed by a project plan, detailing the time, people, and financial resources associated with delivery. (See Chapter 5 for a structured management approach to these matters.)

Note that the model depicted in Exhibit 1 comprehends all of the enterprise's information technology expenditures, including internal labor (IT staff) and external vendor, consulting, and contractor costs. Furthermore, some organizations wisely set aside a reserve each year in anticipation of related but unexpected costs, such as project overruns, emerging technology investments, and IT organization responses to changes in the enterprise's business plans. This IT investment reserve may serve as a contingency fund for both discretionary and nondiscretionary cost overruns, as well as unplanned initiatives. Driven by the number of users and the extent of services, nondiscretionary costs will typically consume at least 60 percent of the annual IT budget and, if not carefully managed, may preclude the opportunity for more strategic IT (project-based) investments. Put another way, the total sum devoted to IT expenditure by the enterprise is rarely elastic. If nondiscretionary costs run out of control, there will be little left for project work. If the business' leadership envisions major new IT investments, these may only come at the expense (if possible) of existing IT services or through enlarging the overall IT allocation.* See Exhibit 1.**

Not surprisingly, the enterprise's leaders usually want it both ways: namely, they expect the IT organization to keep the total cost of its

---

* One of the primary justifications for service level (see Chapter 4) and commitment management (see Chapter 5) is to address the all-too-familiar phenomenon whereby business leaders commit the enterprise to IT investments without fully appreciating the total cost of ownership associated with their choices. Without proper planning, such a course of action can tie the hands of the IT organization for years to come and expose the enterprise to technological obsolescence.

** For an electronic version of the IT internal economy stack and additional guidelines on differentiating SLA work from project work on a cost basis, see *The Hands-On Project Office*, http://www.crcpress.com/e_products/downloads/download.asp?cat_no=AU1991, chpt1~1~internal economy model.

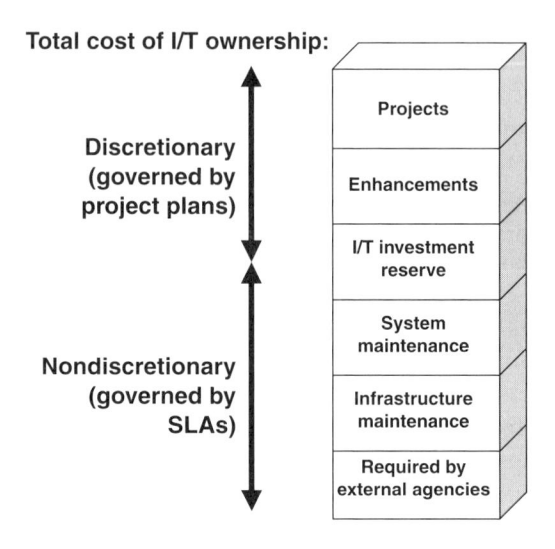

Exhibit 1 The Internal Economy of IT Product and Service Delivery

Total cost of I/T ownership:

Discretionary (governed by project plans):
- Projects
- Enhancements
- I/T investment reserve

Nondiscretionary (governed by SLAs):
- System maintenance
- Infrastructure maintenance
- Required by external agencies

operations steady while taking on new initiatives. For this reason, IT leaders must manage their commitments with great care through a rigorous process of project prioritization, customer expectation management, and resource alignment. To succeed in this endeavor, the information technology organization must keep it simple and keep it collaborative. More specifically, the organization should employ an investment-funding model along the lines mentioned above. It should separate and manage recurring (nondiscretionary) activity through SLAs.[*] Similarly, it should manage projects through a separate but connected commitment synchronization process.[**]

---

[*] SLAs are created through an annual process to address work on existing IT assets, including all nondiscretionary (maintenance and support) IT costs, such as vendor-based software licensing and maintenance fees, and the discretionary costs associated with system enhancements below some threshold amount (e.g., $10,000 per enhancement effort). Typically, SLA work will at times entail the upgrade costs of system/ or Web site hardware and software, as well as internal and external labor costs, license renewals, and so forth.

[**] The project commitment process governs the system development life cycle for a particular project, encompassing all new IT asset project work, as well as those few systems or Web site enhancements that are greater than the SLA threshold project value. Typically, project work will entail the purchase costs of new system or Web site hardware and software, internal and external labor costs, initial product licensing, and so on. Once a project deliverable is in production, its ongoing cost is added to the appropriate SLA for the coming year of service delivery.

Throughout these labors, IT management should employ metrics that measure value to the business and not merely the activity of IT personnel. Last but not least, although IT management should take ownership of the actual technology solutions, it must also ensure that the proper business sponsors, typically line-of-business executive management, take ownership of and responsibility for project delivery and any associated business process changes in partnership with IT counterparts. The remaining sections of this chapter will consider in broad outline the areas of service and project delivery management.

THE THREE PILLARS OF IT DELIVERY

Whatever your organization's existing portfolio of information technology–enabled products and services, your customers view these like a utility. Their expectation is that products and services are always available, stable, reliable, and, of course, affordable. Once established, these products and services become part of the fabric of enterprise operations and only merit notice when they fail or when they cease to meet the user's needs. Thus, IT team delivery revolves around fixing, enhancing, and adding to services, with the existing state of information technology products and services taken as a given. Of course, the IT organization knows better. Keeping current services up and running on a 24/7 basis is no mean task. On the other hand, the perceived value and contribution of an IT team rests largely on the team's ability to address the notion of IT as a utility, even as it struggles to improve, expand, and enhance services. Though seen differently from a technical and management point of view, all of this effort falls under the rubric of IT team delivery.

In the simplest of terms, most information technology organizations provide services in any one of three logical categories: problem resolution, service requests, and projects. First, IT solves customer problems in support of existing products and services. This type of service usually involves a help desk or call center, hardware and software support personnel, and training and documentation services. Problem resolution aims to address the specific IT product and service issues on behalf of the end user as quickly and as painlessly as possible. Second, IT responds to service requests that call either for the extension of existing products and services to a new employee or for the modest expansion and enhancement of an established product or service to existing employees. Here, too, a help desk or call center is often the intake mechanism for service request work, typically complemented by dedicated support and maintenance teams.

Neither problem resolution nor service request efforts individually entail large capital outlays, major changes in platform technologies, or, in most instances, serious commitments of IT personnel. They either fix or build upon what is already there. The customer's expectation is that delivery will be immediate or nearly so. Together, problem resolution and service request delivery, along with ongoing operational costs, constitute the majority of nondiscretionary IT spending. Because this may amount to anywhere from 50 percent to 70 percent of the total IT budget, and because it is typically the first and primary way enterprise and external customers are touched by the IT organization, these services remain job one for most IT shops.

The third category of IT team activity — projects — encompasses the significant expansion of existing products and services or the introduction of new ones. Unlike the aforementioned categories, project work typically requires major capital outlays, a project management infrastructure, the involvement of external technology partner providers, and a long (as opposed to a short or immediate) delivery timeframe. Projects tend to push IT organizations onto the bleeding edge of technology adaptation, but these may be viewed by the team as the most satisfying assignments from a purely technical perspective. Unfortunately, projects also typically encompass both high (but unarticulated) business and technical risks and unbridled expectations among the corporate sponsors of these undertakings. Thus, these discretionary expenditures offer their own set of challenges to IT management and therefore demand a parallel, but somewhat different, approach to disciplined delivery.

As a mental exercise, the reader may wish to consider the unique characteristics of IT problem resolution, service request, and project delivery management as these pertain to the reader's shop and business setting. For example, how does the unique context of your own internal IT economy impact your ability to delivery in these three areas of service? Exhibit 2 shows a matrix that may assist in this effort.

This matrix makes clear that the so-called three-pillars of IT service delivery call upon different types of team skills and resources. In fact, these three pillars often should be treated within IT as three distinct lines of business, procedurally and perhaps organizationally. More often than not, problem resolution services cover familiar territory, involving face-to-face interactions among those who use and those who maintain standard, established IT services. Typically, the problems in question impact the enterprise's ability to service external customers or to execute internal business processes. Solutions will be documented through the help desk and will be familiar to those in IT answering the calls. At times, a problem may escalate into the need to change a major system or infrastructure component, as when the hacking of a server triggers a review of enterprise network security procedures. But more often than not, a quick fix, accompanied by some end-user education, meets the need.

Exhibit 2 IT Services Delivery Matrix

Work Category	Attributes	Typical IT Roles
Problem resolution	IT issues impacting existing products and services Need for quick fixes Need for user training and support Day-to-day overhead costs covered by existing service contracts with vendors and partner providers Just-in-time training Documentation "cheat sheets"	Help desk or call center Access and security control services Desktop support Network support Production services Systems support End-user training and documentation services
Service requests	Installation of new workstations or network connections Installation of patches and upgrades to existing systems Implementation of desktop software Minor software enhancements Day-to-day overhead costs covered by existing service contracts and IT organization operating budgets Just-in-time training Documentation "cheat sheets" and more extensive documentation revisions	Help desk or call center Desktop installation team Network services installation team Production services Systems development and maintenance teams Database administrators End-user training and documentation services
Projects	Major hardware upgrades Major software upgrades Installation of new hardware and software platforms Rollout of major new desktop functionality Implementation of new application systems Enabling major process changes within the enterprise Enterprisewide user training and associated documentation	Business and IT executives Project directors and managers Technology architects Business analysts Network and server services Systems development teams End-user training and documentation services External information technology partners

By contrast, a service request may range widely, from adding a new end user to the network to upgrading or patching a major enterprise software application. To address a service request IT personnel may require detailed knowledge of an application suite or a hardware or software environment. In smaller IT shops, the same group may address problem resolutions and service requests, where tier one tasks are the domain of the call center and help desk and where tier two tasks are assigned to system teams who do maintenance, support, and development on the applications in question.

Project work stands out from the other two pillars of IT service because these assignments call for sizeable resource investments, hold longer time horizons until delivery, and require business process discovery and possibly reengineering efforts. In effect, IT projects aim either to replace one technology with another or to introduce IT to a hitherto manual process. To that end, IT must work closely with the project's business sponsors to expose and analyze the business process and its associated information management needs to bring technology appropriately to bear. This work calls upon different skills, including project management, resource estimating, business analysis, process and data modeling, system prototyping and development, risk management, and so on. Typically, IT organizations structure themselves around the appropriate competencies and technical expertise to deliver on their project commitments. The aforementioned matrix summarizes these attributes and roles. How does this model compare with the realities within the reader's IT organization? What can we learn from this? To assist in drawing these distinctions and relating them to the reader's present situation, let us take our exploration of IT delivery management to the next level of detail.

MANAGING SERVICE DELIVERY

The services delivered by IT to its customers across the enterprise have evolved over time and are in a constant state of flux as both the business needs of the organization and its underlying enabling information technologies evolve.[*] Given this ever-changing landscape and the general inadequacies of the typical lines of communication between IT teams and their customers, much of what is expected from IT is left unsaid and assumed. This is a dangerous situation, inevitably leading to misunderstandings and strained relations all around. The whole point of service level management is for IT to identify customer requirements clearly and proactively, to define IT services in light of those requirements, and to

---

[*] For a particularly comprehensive consideration of this subject, see Rick Sturm, Wayne Morris, and Mary Jander, *Foundations of Service Level Management* (Indianapolis, IN: SAMS, 2000).

articulate performance metrics (i.e., service levels) governing service delivery. Then, the information technology organization should regularly measure and report on IT's performance, working to reinforce the value proposition of IT to its customers.

In taking these steps, IT management will provide its customers with a comprehensive understanding of the ongoing services delivered by the IT organization. Furthermore, service level management establishes a routine for the capture of new service requirements, for the measurement and assessment of current service delivery, and for alerting the customer to emerging IT-enabled business opportunities. In so doing, IT service delivery management will ensure that IT resources are focused on delivering the highest value to the customer and that the customer appreciates the benefits of the products and services so delivered. The guiding principles behind such a process may be summarized as follows:

- Comprehensive — the process must encompass all business relationships and the products and services delivered by IT to its customers
- Rational — the process should follow widely accepted standards of business and professional best practice, including standard system development life cycle (SDLC) methodologies
- Easily understood — the process must be streamlined, uncomplicated, and simple, hence easily accessible to nontechnical participants in the process
- Fair — through this process, the customer will understand that he or she pays for the actual product or service as delivered; cost and service level standards should be benchmarked and then measured against other best-in-class providers
- Easily maintained — the process should be rationalized and largely paperless, modeled each year on prior-year actuals, and subsequently adjusted to reflect changes in the business environment
- Auditable — to win overall customer acceptance of the process, key measures must be in place and routinely employed to assess the quality of IT products, services, and processes

The components of the IT service delivery management process include the comprehensive mapping of all IT services against the enterprise communities that consume those services. It also includes service standards and performance metrics (including an explicit process for problem resolution), the establishment and assignment of IT customer relationship executives (CREs) to manage individual customer group relations, a formal SLA for each constituency, and a process for measuring and reporting on service delivery. Let us consider each of these in turn.

As a first step in engineering an IT service level management process, IT management must segment its customer base and conceptually align IT services by customer. If the information technology organization already works in a business environment where its services are billed out to recover costs, this task may be easily accomplished. Indeed, in all likelihood such an organization already has SLAs in place for each of its customer constituencies. For most enterprises, however, the IT organization has grown along with the rest of the business and without any formal contractual structure between those providing services and those being served.[*] If your enterprise falls in the latter category, begin your assessment of the situation by employing an enterprise organizational chart to map your IT service delivery against that structure. As you do, ask the following questions:

- What IT services apply to the entire enterprise and who sponsors (i.e., pays for or owns the outcome of) these services?
- What IT services apply only to particular business units or departments and who sponsors (i.e., pays for or owns the outcome of) these services?
- Who are the business unit liaisons with IT concerning these services, and who are their IT organization counterparts?
- How do the business unit and IT measure successful delivery of the services in question? How is customer satisfaction measured?
- How does IT report its results to its customers?
- What information technology services does IT sponsor on its own initiative, without any ownership by the business side of the house?

Obviously, the responses to these questions will vary greatly from one organization to another and may in fact vary within an organization, depending on the nature and history of working relationships between IT and the constituencies it serves. Nevertheless, it should be possible to assign every service IT performs to a particular customer group or groups, even if that group is the enterprise as a whole. Identifying an appropriate sponsor may be more difficult, but in general, the most senior executive who funds the service or is held accountable for the underlying business enabled by that service is its sponsor. If too many services are owned by

---

[*] Unless the enterprise's leadership has chosen to operate IT as a separate entity with its own profit-and-loss statement, the author would recommend against the establishment of a charge-back or transfer pricing system between IT and its customers. Instead, the author recommends that the business and IT leaderships agree jointly on the organization's overall IT funding level and that IT manage those funds in line with the service level and project commitment management processes and performance metrics outlined in this book.

your IT executive team rather than line-of-business leaders, IT may have a fundamental alignment problem. Think broadly when making your categorizations. If a service has value to the customer, some customer must own it; if no customer comes forward, why is IT delivering that service in the first place? In doing such analysis for employers and clients, the author has regularly uncovered services that the enterprise no longer requires. These services were subsequently eliminated, freeing IT resources for work of greater customer value.

In concluding this analysis, the IT team will have identified and assigned all of its services (nondiscretionary work) to discrete stakeholder constituencies. This body of information may now serve as the basis for creating SLAs for each customer group. The purpose of the SLA is to identify, in terms that the customer will appreciate, the products and services that IT brings to that group. The purpose of the SLA, however, goes well beyond a listing of services. First and foremost, it is a tool for communicating vital information to key constituents about how they may most effectively interact with the IT organization. Typically, the document includes contact names, phone numbers, and e-mail addresses. Second, an SLA also helps shape customer expectations in two different but important ways. On the one hand, an SLA identifies customer responsibilities in dealing with IT. For example, it may spell out the right way to call in a problem ticket or a request for a system enhancement. On the other hand, it defines IT performance metrics for standard services and for the resolution of problems and customer inquiries. Last but not least, a standard SLA compiles all the services and service levels that IT has committed to deliver to that particular customer, reinforcing the value proposition between that business entity and the information technology organization.

SLAs can take on any number of forms.* Whatever form you chose, ensure that it is as simple and brief a document as possible. Avoid technical jargon and legalese, and be sensitive to the standard business practices of the greater enterprise within which your IT organization operates.** Most of all, write your SLAs from your customer's perspective, focusing on what is important to the customer. State in plain English what services the customer receives from you, the performance metrics for which IT is

---

* See Sturm, Morris, and Jander, pp. 189–196.
** For example, an SLA that resembles a formal business contract is appropriate and necessary for a multi-operating unit enterprise in which each line of business runs on its own P&L and must be charged back for the IT services that it consumes. On the other hand, such an SLA would only confuse and frustrate the executives of an institution of higher education, who are unaccustomed to formal and rigorous modes of business communication and operate more collegially in an informal context of mutual trust and respect.

accountable, and what to do when things break down or go wrong. A more detailed discussion of the SLA may be found in Chapter 4. Whatever the final form of your SLA, prepare SLAs for your entire customer base at one time to ensure that, in total, they address all IT services as delivered and their associated consumption of IT resources.

Your next step is to assign a customer relationship executive (CRE) to each SLA account. The CRE serves as a primary point of contact between customer executive management and the IT organization. In this role, the CRE will meet with his or her executive sponsors for an initial review of that business unit's SLA and thereafter on a regular basis to assess IT performance against the metrics identified in the SLA. Where IT delivers a body of services that apply across the enterprise, you might consider creating a single, community SLA that applies to all and then brief addenda that list the unique systems and services pertaining to particular customer groups. Whatever the formal structure of these documents, the real benefit of the process comes from the meetings in which the CRE can reinforce the value of IT to the customer, listen to and help address IT delivery and performance problems, learn of emerging customer requirements, and share ideas about opportunities for further collaboration between the customer and IT. Within IT, the CRE will act as the advocate and liaison for, and as the accountable executive partner to, his or her assigned business unit in strategic and operational matters.

CREs must be chosen with care. They should be good listeners and communicators. For the customer in question, they must have a comprehensive understanding of what IT currently delivers and what information technology may afford. Although they need not be experts in all aspects of the business conducted by the customer, they must at least have a working knowledge of that business, its nomenclature, and the roles and responsibilities of those working in that operating unit. Among the many skills that a good CRE must possess is the ability to translate business problems into technical requirements and technical solutions into easily understood narratives that the customer can appreciate. The CRE also must be a negotiator, helping the customer to appreciate the breadth of that business unit's portfolio of IT services and projects, within the limitations of the resources available to serve those needs. At some times, such conversations may require that the customer choose among existing options, and at others, defer a request for IT services to a more opportune occasion.

IT customers will appreciate immediately the value of the CRE, because this role offers them a single channel for addressing their strategic and tactical IT needs. Over time, this arrangement may lead to abuses, such as customers who approach the CRE with problems more appropriately directed to the IT help desk or an IT line service provider. When this

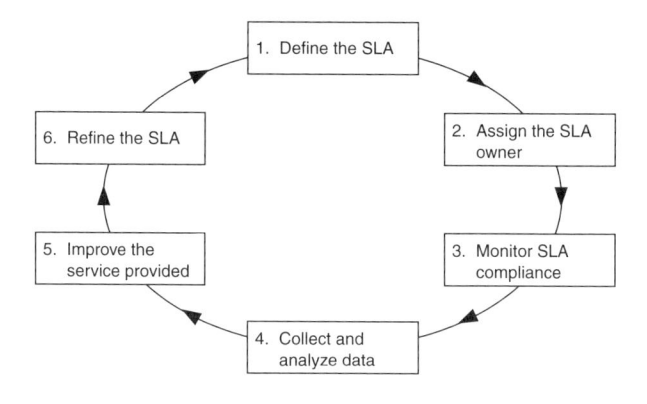

The SLA management process diagram shows a cyclical flow:
1. Define the SLA
2. Assign the SLA owner
3. Monitor SLA compliance
4. Collect and analyze data
5. Improve the service provided
6. Refine the SLA

Exhibit 3 The SLA Management Process

occurs, the CRE should take the information, indicate that he or she will forward it to the proper person in IT, and suggest that the customer go directly to the appropriate IT party in the future. Here, the CRE functions as a facilitator of best practices in the working relationship between a given customer group and the information technology organization. He or she can also note weaknesses in IT procedures and work behind the scenes with IT colleagues on process improvements. The greatest value of the CRE, however, is to act as a human link to a key IT customer constituency, managing customer expectations while keeping IT focused on the quality delivery of its commitments to that group.

This effort is iterative: collecting and processing data, meeting with customers and IT service providers, listening, communicating, and educating. As depicted in Exhibit 3, SLA process management operates on an annual calendar. The CRE reports to the customer on a monthly or more frequent basis and brings back data to refine and clarify existing service delivery, as well as the service objectives for the coming year.

Again, the key to success for the CRE and the SLA process is attention to detail. Customers hate surprises. They can understand that IT will fail from time to time; after all, what technical or, for that matter, human process components do not? What matters is keeping the customer apprised of what is going on and what IT is doing to address performance difficulties. If the customer remains well informed, much of the tension between IT and those it serves will dissipate as trusting business relationships mature. By respecting the cycle of service delivery and by communicating regularly and with full disclosure, the overall process will pay off for you and your IT team.

The service level management process will ensure the proper alignment between customer needs and expectations on the one hand, and IT resources on the other. The process clearly defines roles and responsibilities, leaves little unsaid, and keeps the doors of communication and understanding open on both sides of IT service delivery. From the standpoint of IT leadership, the SLA process offers the added benefit of maintaining a current listing of IT service commitments, thus filling in the nondiscretionary layers of IT's internal economy model. Whatever resources remain may be devoted to project work and applied research into new, IT-enabled business opportunities. Bear in mind that this is a dynamic model. As the base of embedded IT services grows, a greater portion of IT resources will fall within the sphere of nondiscretionary activity, limiting project work. The only way to break free from these circumstances is to curtail existing services or to broaden the overall base of IT resources.

In any event, the service level management process will provide most of the information that business and IT leaders need to make informed decisions concerning the scope, quality, and value of day-to-day IT service delivery. Chapter 4 contains a more detailed consideration of service delivery management tools and techniques. With the service delivery side of the house clarified, let us now turn to the discretionary side of IT investment: project delivery. IT project work calls for a body of management practices complementary to those on the service side if such efforts are to succeed.

MANAGING PROJECT COMMITMENTS

To put it simply, any IT activity that is not covered through an SLA is a project that must be assigned IT discretionary resources. The author takes it as a given that the enterprise employs a planning process of some type whereby IT projects are identified and prioritized. (See Chapter 3 for a detailed consideration of overall IT organization planning, prioritization, and resource alignment.) IT is then asked to proceed with this list, in line with available resources. Once the business has sorted out what it plans to do for the year, the information technology organization must have in place its own process for governing the system development life cycle (SDLC) for each approved project. This process will address all new IT work, as well as those few system or Web site enhancements that are greater than the SLA cost threshold for such work. Typically, project work will entail the purchase costs of new system or Web site hardware and software, as well as internal and external labor costs, initial product licensing, and so forth. It is also standard practice to treat as project costs maintenance and support costs incurred during the life of the project and through the remainder of the first fiscal year of its delivery. Once a project deliverable is in production, its ongoing cost is added to the appropriate SLA for the coming year of service delivery.

All of this sounds straightforward. Unfortunately, IT organizations find it much easier to manage and deliver routine, ongoing SLA services than to execute projects. The underlying reasons for this state of affairs may not be obvious, but they are easily summarized. Services are predictable events, easily metered, with which IT personnel and their customers have considerable experience and a reasonably firm set of expectations. More often than not, a single IT team oversees day-to-day service delivery in a particular area of technology (e.g., network operations, Internet services, e-mail services, security administration, and so forth). Projects, on the other hand, typically explore new territory and require an IT team to work on an emerging, dynamic, and not necessarily well articulated set of customer requirements. Furthermore, most projects are by definition cross-functional, calling on expertise from across the IT organization and that of its business unit customers. When many hands are involved and when the project definition remains unclear, the risk of error, scrap, and rework are sure to follow. These are the risks that the project commitment management process must mitigate.

From the outset, it must be stressed that, like the SLA process, the effort and rigor of managing project commitments will vary from one organization to another and from one project to the next. The pages that follow provide a framework for informed decision-making by the enterprise's business and IT leaders as they define, prioritize, shape, and deliver IT projects. The desire to pursue best practices must be balanced with the real-world needs of delivery within a particular business environment. (See Chapter 5 for a more detailed consideration of project management, measurement, and reporting tools.)

As a first step, the enterprise's business leadership will work with IT to identify appropriate project work. Any efforts that appropriately fall under existing SLA agreements should be addressed through the resources already allocated as part of nondiscretionary IT funding for that work. Next, each CRE will work with his or her executive sponsor(s) to define and shape potential project assignments for the coming year. Although each CRE will assist in formulating and prioritizing these project lists, he or she must make it clear that this data-gathering activity in no way commits IT. Instead, the CREs will bring these requests back to IT executive management, who will in turn consolidate and rationalize these requests into an IT project portfolio for the review and approval of the enterprise's leadership.[*] This portfolio presentation should indicate synergies and

---

[*] It is essential that the business and not IT rule on project priorities. However, this does not abdicate the IT team's responsibility to consolidate and leverage IT requests that, in its view, bring the greatest benefit to the enterprise, and to identify infrastructure and other IT-enabling investments that are a necessary foundation for business-enabling IT projects.

dependencies between projects, the relative merits and benefits of each proposal, and the approximate level of investment required and risks associated with each proposed undertaking.

With this information in hand and as part of the annual budgeting and planning process, the enterprise's business and IT leaders will meet to prioritize the list and to commit in principle to those projects that have survived this initial review. Typically, all enterprise-level projects emerging from and funded by this process are of the highest priority in terms of delivery and resource allocations. If additional resources are available, business unit–specific projects may be considered in terms of their relative value to the enterprise. In the for-profit sector, enterprises will define a return-on-investment (ROI) hurdle rate for this part of the process, balancing line-of-business information technology needs against overall enterprise IT needs. In many instances, business units may receive approval to proceed with their own IT projects as long as they can fund them and as long as internal IT has the bandwidth to handle the additional work. Invariably, unforeseen circumstances and business opportunities will necessitate revisiting the priority list. Some projects may be deferred and others dropped, in favor of more pressing or promising IT investments.

Similarly, as the IT team and its business partners work through the development life cycle on particular projects, they will find that their original assumptions are no longer valid, requiring the resizing, rescheduling, redefinition, or elimination of these projects. The key to success here is the employment of an initial, rigorous project-scoping effort, coupled with a comprehensive project life-cycle management process that ensures regular decision points early in the project's design, development, and implementation phases. Once a project is properly scoped and enters the pipeline, the IT project director,[*] working in collaboration with his or her working client(s) and supported by an IT project manager,[**] will create a commitment document and a project plan (both to be discussed later), reflecting detailed project commitments and resource allocations. The IT CRE will then monitor the project team's overall compliance with the plan, reporting to the customer on a regular basis.

---

[*] The project director is the IT party responsible for project delivery and the overall coordination of internal and external information technology resources. He or she will work hand-in-hand with the customer working clients to ensure that project deliverables are in keeping with the customer's requirements.

[**] The IT project manager is staff to the IT project director. This support person will develop and maintain project commitment documents and plans, facilitate and coordinate project activities, carry out business process analysis, prepare project status reports, manage project meetings, record and issue meeting minutes, and perform many other tasks as required to ensure successful project delivery.

Initial project scoping is key to the subsequent steps in the project management process. All too often, projects are pursued without a clear understanding of the associated risks and resource commitments. Neither the project's working clients nor its IT participants may understand their respective roles and responsibilities. Operating assumptions are left undocumented, and the handoffs and dependencies among players remain unclear. Without sufficient information along these lines, IT efforts so undertaken will usually end in severe disappointment. To avoid such unhappy results, IT project teams should embrace a commitment process that ensures a well informed basis for action.

From the outset, no project should proceed without an executive (business) sponsor and the assignment of at least one working client. The executive sponsor's role is to ensure the financial and political support to see the project through. He or she owns the result and is therefore the project's most senior advocate. The sponsor's designated working clients are those folks from the business side of the house who will work hand-in-hand with IT to ensure a satisfactory delivery of the project. Without this level of commitment from the business, no project should proceed. It is best that the sponsor appoint appropriate working clients, making it clear that they have as much at stake in the success of the project as their IT counterparts. On the other hand, avoid assigning working clients as project directors or managers.* Although these folks have the business knowledge essential to the project's success, they often lack the requisite skills to manage the coordination of IT resources and delivery. If the project in question happens to be sponsored by IT itself, then the chief IT executive will serve as sponsor and the IT manager who will own the system or service once it is in production will serve as the working client. Although it is assumed that the project is funded, the commitment document should indicate the project's recognized priority. The overall process must also capture information concerning the project's scope, the definition of delivery and associated customer satisfaction metrics, project risks, the roles and responsibilities of the project team, and so forth.

Unfortunately, many of the skills required to draw out this information from IT customers do not typically reside in IT line managers. These people focus, day in and day out, on the 24/7 availability of IT and may

---

* Some organizations will allow working clients to serve as IT project directors and even project managers. In the author's view, this is a mistake. Although the working client is essential to any IT project's success, contributing system requirements and business process expertise to the effort, very few working clients have experience in leading multitier IT projects, especially those involving outside technical contractors and consultants. Leave this work to an appropriately skilled IT manager, allowing working clients to contribute where they add greatest value in articulating and clarifying business requirements and system performance standards for the IT team.

lack both the intimate business process knowledge and the people-management skills required to succeed in project delivery. Chapter 2 will address this subject in more detail as it justifies the creation of a formal or virtual project management office (PMO) within the information technology organization. For now, it suffices to close with the observation that, like the SLA process, project management requires its own unique portfolio of skills and tools. In particular, customer (i.e., sponsor and working client) management must complement process, risk, and resource managements. In terms of the IT organization, the CRE facilitates, coordinates, and communicates between the business and IT teams. But the CRE role is not designed for day-to-day oversight of detailed project implementation. This work rests with project directors, project managers, business analysts, and other IT personnel assigned to each project. As Chapter 2 shows, the PMO may serve as the ideal instrument for ensuring the success of these more granular efforts. As part of this work, each project team must agree on benchmarks for success: what is to be measured, how it is to be reported, and when communication is to take place. The simplest of formulas works here: report regularly, use the customer's language and metrics, and always tell it like it is.

IT METRICS AND REPORTING TOOLS

Given all of the work that a typical IT organization is expected to deliver each year, it is easy to see how even major commitments may be overlooked or misinterpreted in the furious effort to get things done. To avoid this pitfall, IT must clarify its commitments to all concerned. Prior sections of this chapter have shown how this may be done for both ongoing service delivery and project implementations. Next, IT management must ensure compliance with its commitments once made. Here again, the aforementioned processes keep the team appropriately focused. Service level management requires the integration of performance metrics into each SLA, while the ongoing project management process forces the team to relate actual accomplishments to its project plan. During the regular visits of the CRE with his or her customers, service level and project delivery may be raised with the customer to assess his or her current satisfaction with IT performance.

Although each of these activities is important in cementing and maintaining a good working relationship with individual customers, a more comprehensive view is required of how IT services and projects relate to one another. For its part, IT executive management needs a more aggregate view of the unit's performance, especially in larger organizations. To this end, the author has relied on a single, integrated reporting process, called the monthly operations report, to capture key

IT accomplishments and performance metrics. Chapter 4 and Chapter 5 will comment on the operations report format in detail. What follows is an introductory overview.

At its name implies, the monthly operations report is a regularly scheduled activity. The document is entirely customer-focused and therefore aligns with both the service level management and project commitment processes. However, the monthly operations report is designed primarily to serve the needs of IT management, keeping customer delivery at the forefront of team's attention and holding IT personnel accountable for their commitments. The report reflects qualitative information from each IT service delivery unit (e.g., help desk, training center, network operations, production services, and so forth) concerning accomplishments and issues during the course of the month. Each accomplishment must be aligned with a customer value, as articulated in SLAs and project commitment documents, if it is to be listed as a deliverable. Each issue must address who was impacted by the product or service failure, as well as how it was resolved. Next, the report lists quantitative performance data, such as system availability, system response time, problem tickets closed, training classes offered, and the like. Note that some of these data points measure activity rather than results and must be balanced with customer satisfaction metrics to be truly meaningful.

To that end, the author has developed and deployed a low-cost system for collecting customer feedback. This simple surveying process is guided by the following operating principles:

- First, the process must require no more than two minutes of an individual customer's time
- Second, it must be conducted via the phone or face-to-face but never via paper forms or e-mail, although recently available Web-based surveying tools are much less obtrusive and could serve as a surrogate for actual human interaction
- Third, it must employ measures of customer satisfaction rather than IT activity
- Fourth, at the very least, it must scientifically sample IT customer populations
- Fifth, it must be carried out in a consistent manner on a regular basis

Guided by these principles, the author's own project management office team has implemented effective survey tools for many types of IT service.

To initiate these processes, the team has drawn randomly from the help desk customer database, where requests for service and problem tickets are recorded. A single staff member spends the first few days of each month calling customers, employing the appropriate survey scripts.

Results are captured in a simple database and then consolidated for the report. These customer satisfaction scores are also tracked longitudinally. Initially, my IT colleagues were skeptical, but now that they appreciate the objectivity of the process, they value the useful information that it generates. For their part, our customers have given us high marks for launching a process that tracks and reports publicly on their satisfaction (or not!) with our services. All the summary data appears in the monthly operations report."*

Project delivery is a little more complicated to capture on a monthly basis because projects do not necessarily lend themselves to quantitative measures or to regular surveying of customer satisfaction. Nevertheless, the operations report contains two sets of documents that IT management finds useful. The first is a project master schedule that lists all current and pending IT projects alphabetically by title, indicating IT ownership (i.e., project director), and interproject dependencies. The schedule also shows the duration of each project and its status: white for completed, green for on schedule, yellow for in trouble but under control, red for in trouble, and purple for pending. Thus, within a few pages, the IT leadership can see all of the discretionary work under way at any given time, what is in trouble, where the bottlenecks are, and who is over committed.

The presentation is simple and visual. Within the operations report, each project has its own scorecard, a single-page representation of that project's status. Like everything else in the report, the scorecard is a monthly snapshot that includes a brief description of the project and its value to the customer, a list of customer and project team participants, this month's accomplishments and issues, a schematic project plan, and a Gantt chart of the current project phase's activities. Like the master schedule, scorecards are coded white, green, yellow, red, or purple as appropriate. (See Chapter 5 for more details and *The Hands-On Project Office* at http://www.crc-press.com/e_products/downloads/download.asp?cat_no=AU1991 for template examples.)**

---

* This process for collecting customer satisfaction data operated between 2000 and 2003 at Northeastern University, where it was subsequently replaced by a Web-enabled survey process built with the eSurveyor tool. With the introduction of eSurveyor, integrated with the Information Services Division's Remedy problem-tracking system, the PMO now has an automated way to reach all customers, obviating the need to sample trouble ticket callers.

** Also see *The Hands-On Project Office*, http://www.crcpress.com/e_products/downloads/download.asp?cat_no=AU1991, chpt4~6~monthly service delivery report~template, chpt4~7~monthly service delivery report~example, chpt5~15~project scorecard~template, chpt5~17~monthly project status report~template, and chpt5~18~monthly project status report~example.

The monthly operations report is reviewed each month in an open forum by the IT executive team. Other IT personnel are welcome to attend. Within a two- to three-hour block, the entire IT leadership team has a clear picture of the status and health of all existing IT commitments. Follow-up items raised in the review meeting are recorded and appear in the next version of the report. The document itself is distributed to the entire IT organization via the unit's intranet. As appropriate, sections from the report, as well as individual project scorecards, are shared by the unit's CREs with their respective customers. In brief, the process keeps accomplishments and problems visible and everyone on their toes. Bear in mind, the focus of this process is continuous improvement and the pursuit of excellence in customer delivery. Blame is never individually assessed, because the entire IT team is held accountable for the success of the whole. On the other hand, it is marvelous to observe how much more conscientious IT service and project delivery managers are when they know that they must face their peers each month to report on their progress — or the lack thereof.

In a world of growing complexity, where time is of the essence and the resources required to deploy IT effectively remain constrained, the frameworks and simple tools outlined in this chapter have proved useful to me, and, if adapted to the reader's own work environment, should serve him or her well. The underlying principles for these efforts are simple, and the practices themselves are commonsensical. These remain the keys to success:

- A solid focus on customer value
- Persistence in the use of standard, rigorously defined but creatively adapted and flexibly executed processes
- Quality and continuous communication
- A true commitment to collaborative work

The reader will do well to bear this in mind as we dig deeper into the processes mentioned in this chapter.

On the other hand, the picture painted thus far should raise concerns among some readers. The author advocates a series of internal IT management practices that perhaps enjoy no precedent within the reader's IT organization. In short, I am asking you to add still more to your overhead costs. This is true. The recommendations in this book require an investment of time and effort on the part of the IT team. You may not have the right skills on hand to get this particular set of activities accomplished. Furthermore, you may not, in your view, have the bandwidth to add these responsibilities to existing organizational roles. Yet ask yourself the following questions:

- What is the quality of your relations with your customers today, and what is the quality of communication between your line-of-business sponsors and IT?
- How much scrap and rework do you pursue annually?
- How many projects fail?
- What is the enterprise's view of IT service delivery?
- Is the information technology organization getting the resources it needs to satisfy customer requirements?

It should be clear where I am headed. The cost of operating as I suggest will have an impact on IT resources, but these demands pale in comparison to the risks and costs that you and your team face through ineffective communication and tarnished service and project delivery. The next chapter addresses how IT should configure for successful delivery management, as well as the tools to justify such an investment in people and process change.

2

THE PROJECT MANAGEMENT OFFICE BUSINESS MODEL

INTRODUCTION: REVISITING THE IT ORGANIZATION

The opening sections of this book have set the stage for a review of the information technology organization and the arguments for creating a project management office (PMO). In preparing this ground, the author has stressed the central importance of service and project delivery to the value proposition of any information technology organization. But achieving success in these areas is a challenge for most IT teams who are obliged to balance the at-times conflicting and rapidly changing needs and demands of the parent enterprise with responses constrained by limited financial and human resources. To assist the reader in dealing with these complexities, the author offers a toolkit. But this still begs the question, "Who will wield the tools?" The answer will depend on the organizational context of the reader's IT business, the strengths and weaknesses of its players, and the particulars on its to-do list. This chapter explores the various dimensions of the IT organization, service and project delivery roles and responsibilities, and how the project management office might be positioned within this mix.

I have already suggested perspectives, processes, and tools to better position information technology investments within the internal economy of the enterprise. Similarly, I have provided an overview of IT organization day-to-day service operations and project delivery. Besides the actual planning process and its associated documentation, I have recommended the creation of a customer relationship executive (CRE) role, as well as those of the project director, the project manager and the project management office. Introducing these concepts calls for process changes within the IT organization itself, as well as fairly significant alterations in the ways that IT interacts and communicates with its customers and partner providers.

This is a lot to do, and much of it will come as an added burden to already overtaxed IT management. Furthermore, my integrated management perspective, which is focused on the timely delivery of commitments, does not readily fit within the typical roles and responsibilities of an IT operating unit. Bred by a need for attention to detail and the care and feeding of a 24/7 services utility, the mindset of the typical IT team is linear, transaction-based, and heavily siloed. It is therefore not surprising to find that some of these teams are severely challenged in their attempts to function within the more global operating parameters outlined in the opening sections of this book. Established IT structures, business processes, and ways of thinking work to thwart the development of the very cross-functional, business-oriented capabilities and perspectives required to balance this linear, transaction-based mindset.

Consider for a moment a typical but simplified IT reporting structure and how this design reinforces the behavioral characteristics just mentioned. The average IT organization comprises several groupings of professionals clustered around like activities and technologies. There will be an infrastructure team whose responsibilities encompass data-center operations, network administration, server and storage management, production services, and so forth. In many instances, information security, disaster recovery, and contingency planning also fall under this operations group. Another subset of IT personnel may be assigned to customer support, including such services as help desk and call center operations, desktop and audio/visual support, and end-user training and documentation. To complement these teams, IT organizations will also possess a systems organization, in which specially skilled personnel develop,[*] maintain, enhance, and support application software for the business. The systems staff may be organized in any number of ways. In large IT shops, each systems complex (e.g., the applications for finance, manufacturing, distribution and logistics, and so forth) may have its own team, further separated into those who

---

[*] Increasingly, many enterprises are turning to packaged software provided by third-party vendors rather than in-house, home-grown software applications. Although there are always exceptions to this rule and although the uniqueness of certain businesses will sometimes require custom-coded systems, many IT teams today buy their software off the shelf and then work with their vendors (i.e., external partner providers) to adapt these products to the needs of their line-of-business customers.

The Project Management Office Business Model ▪ 27

```
                    ┌─────────────────┐
                    │ Chief Information│
                    │     Officer     │
                    └────────┬────────┘
        ┌────────────┬───────┴───────┬────────────────┐
   ┌─────────┐  ┌─────────┐    ┌─────────┐   ┌──────────────┐
   │Operations│  │ Customer│    │ Systems │   │Administration│
   │         │  │ Services│    │         │   │ and Finance  │
   └────┬────┘  └────┬────┘    └────┬────┘   └──────────────┘
        │            │              │
   ┌────┴────┐  ┌────┴────┐    ┌────┴────┐
   │Data Center│ │Call Center│  │Application│
   │ Services │ │and Help Desk│ │  Team A  │
   └────┬────┘  └────┬────┘    └────┬────┘
   ┌────┴────┐  ┌────┴────┐    ┌────┴────┐
   │ Network │  │ Desktop │    │Application│
   │ Services│  │ Support │    │  Team B  │
   └────┬────┘  └────┬────┘    └────┬────┘
   ┌────┴────┐  ┌────┴────┐    ┌────┴────┐
   │Production│ │  User   │    │   Web   │
   │ Services │ │ Training│    │ Services│
   └────┬────┘  └────┬────┘    └────┬────┘
   ┌────┴────┐  ┌────┴────┐    ┌────┴────┐
   │   IT    │  │Documentation│ │  Data   │
   │Security │  │             │ │ Services│
   └─────────┘  └─────────┘    └─────────┘
```

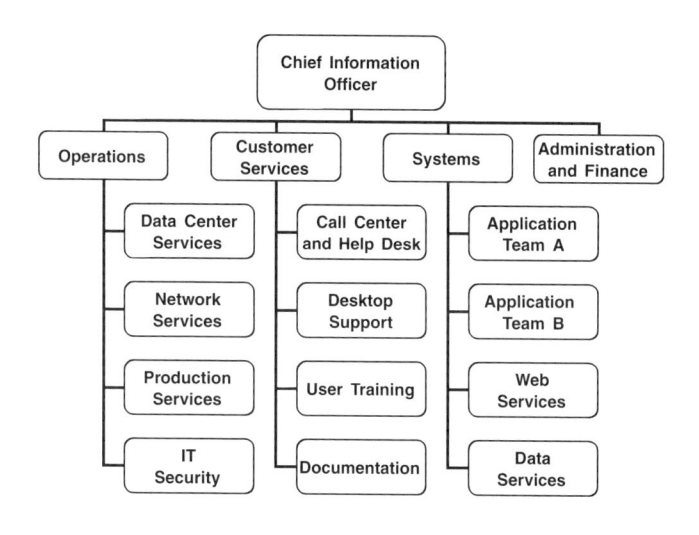

Exhibit 1 Typical IT Organization Design

develop versus those who maintain and support products.[*] Specialized disciplines, such as data management and Web services, may have their own development and maintenance teams. Quality assurance and release management may be embedded in these systems teams, or they may stand as separate IT departments. Lastly, most IT organizations will have some sort of administration and finance function to oversee budgeting, purchasing, human resources management, administrative support, staff development and training, and the like."[**] See Exhibit 1.

It is easy to understand why IT organizations have evolved into the organism described here. The field of information technology is extremely complex, requiring high levels of specialization. To achieve the critical mass of expertise required to run and service an enterprise's technology complex, IT teams usually align themselves according to their technical competencies around particular hardware or software products, information security, cus-

---

[*] No matter the size of your IT organization, keep application development, maintenance, and support for a given product suite within the same team and rotate your people through these various roles and responsibilities. The two primary benefits to this approach are that developers will take greater care in their work if they know they will be saddled with ongoing maintenance of the product, and that sharing the more creative aspects of the work among team members means these same team members will feel less dissatisfaction when they are occasionally assigned more mundane maintenance and support work.

[**] See *The Hands-On Project Office*, http://www.crcpress.com/e_products/downloads/download.asp?cat_no=AU1991, chpt2~1~IT Organization~model, for adaptable electronic versions of the IT organization model.

tomer servicing, and so on. This arrangement makes a lot of sense and generally serves the needs of both the IT organization and its customers.

Unfortunately, there are several less positive aspects to this operational model. First and foremost, it breeds a siloed mentality around service delivery. When organized around particular technologies, IT team members naturally tend to focus on their own technical and service needs and those of their immediate customers, without considering the implications of their actions on other participants along the IT value chain. For example, a network services team may work rapidly to restore a server that has crashed but may fail to communicate the server's failure or its return to service to those application teams that need to take follow-on actions to restore their respective applications. Second, and as a corollary to the first point, highly specialized teams lose sight of what is happening elsewhere in the organization, leading to disconnects in the coordinated planning of deliverables, in the purchase of product, and in the installation of new services. Indeed, how many times have you observed an instance in which a project launch team failed to communicate fundamental hardware, infrastructure, and even facilities requirements to the upstream providers of those capabilities?

Third, when IT personnel work in organizational and mental silos and when a full appreciation of the value chain of IT team activities is missing, IT products and services break more readily. At the very least, they may come to market without having everything necessary in place to ensure customer satisfaction with the deliverables, such as end-user training, documentation, and call center support. Fourth, when things go bad in a siloed environment, finger pointing is rampant. Each work group is confident that it has delivered as required and blames its internal (or external) partner providers for product or service delivery problems. Fifth, given this context, it is difficult to measure and report on end-to-end customer delivery or even to coordinate communications between IT and the customer. From the customer's standpoint, this symptom manifests itself in the need to make many calls across the IT organization to find the right person to address a particular performance issue. Lastly, without a holistic view of IT deliverables it is difficult, if not impossible, to forecast customer needs, to proactively address problems, to allocate resources, and to coordinate IT responses to crises.

How do these aspects of IT organizational design manifest themselves in the real world of IT operations and delivery? Consider the following scenarios. Do they resonate? Have you experienced them yourself, and if so, why?

- Mainframe or a server bank goes down overnight, and the infrastructure team comes in to restore services. The team gets all the computer hardware back online but fails to contact its systems colleagues, who must in turn reboot and reconfigure their applications so that IT's customers can access the services that matter to them. As a result, IT service levels to end customers are severely impacted.

- Through heroic effort, a systems development team brings its product to market on time, only to realize that the team has neglected to purchase enough server hardware for the production instance of the application. Operations indicates that it will now take six to eight weeks to bring the necessary hardware online. The product launch is delayed; systems and operations are angry with one another.
- An application is launched and announced to the community, but the help desk is never notified before the release. End users access the system and experience problems. The help desk is inundated with calls, but without the training and documentation to address user needs, help desk personnel are themselves helpless in this situation. In the end, all of IT looks ill prepared and unprofessional. Again, IT teams play the blame game.
- Customers contact the call center with their IT problem and service requests. Call center personnel log these issues into the organization's problem management system. Unfortunately, there is no consistent practice for problem ticket handoffs or overall process monitoring. As a result, tickets sit in abeyance for weeks on end, leaving customer issues unresolved and customers highly dissatisfied.
- Although the IT leadership has committed to several large and complex projects for delivery within the current fiscal year, the team is understaffed and underskilled to deliver on these commitments. To close this gap, the project teams would like to proceed with the hiring of temporary contractors and the outsourcing of some work. To proceed, however, they must run any contractual arrangements by the legal department and follow the business's hiring and affirmative action procedures. They face months of delays before coming up to appropriate staffing levels. Thus out of the gate, their approved project is seriously behind schedule. The IT leadership is unhappy with the delivery team for underperforming, and the team is unhappy with management for overcommitting.
- IT proudly delivers a new system only to discover that the product was never properly tested and that it cannot scale once placed in a real-world production setting. The team must go to the sponsor for more money and time. Again, IT looks bad in the eyes of the customer.
- IT development team schedules a launch meeting with a key customer. At the final product review, the sponsor and the working clients for the project indicate that the product as delivered does not meet their minimal requirements. Although the customer blames IT for not listening, the fact is that there was no change management process in place. The project must now be restarted with additional resources.

Such scenarios occur commonly within some IT shops, due in part to the lack of coordination and communication among operating unit silos and in part to a breakdown in understanding between the customer and the IT team. To compensate for these process shortfalls, many resort to last-minute heroics to salvage bad situations and turn around misdirected projects. Effective in the short term, such efforts do little to address the dysfunctional behaviors and processes within the IT team and the larger organization. In the long term, they further tarnish the reputation of IT, diminish its perceived value, and contribute to the overburdening of personnel. Last but not least, such experiences waste enterprise resources and make it more difficult to win executive support for the next major IT undertaking.

To close this performance gap, IT organizations could reorganize along cross-functional lines. Unfortunately, this approach carries with it other shortcomings, including the fragmentation of centers of technical competency and excellence and the loss of in-depth staff expertise in critical areas of delivery. Practically speaking, the reason most IT organizations structure themselves around areas of competency (e.g., hardware, software, networking, customer service, and so on) is that this approach makes operational sense. To improve overall results, however, this model requires an added function, one that focuses first and foremost on the process side of IT delivery. This service layer would also concern itself with customer relationship management, the documentation of IT best practices, performance measurement and reporting, and IT staff training and development: in short, a layer that focuses on all of those vital support services that would otherwise fall below the radar screen of the typical, siloed information technology organization.

The Hands-On Project Office therefore recommends an approach whereby a project management office (PMO) is layered into the existing IT organization. As envisioned, the PMO would operate independently of other IT operating units, report directly to the chief information officer (CIO) or some other senior IT executive, and offer a series of support services across IT. See Exhibit 2.

The primary focus of the PMO is to oversee the internal coordination of IT service and project delivery. To do this, the PMO would draw upon IT subject-matter experts operating within their respective operational silos. Serving above the fray, the PMO team remains divorced from the parochial interests of individual IT teams while encouraging common best practices, an end-to-end delivery process perspective, and overall better communication within IT teams and between IT teams and their customers and partner providers. To be sure, this is a tough assignment that must balance objectivity and discernment in service and project delivery with respect for the traditions and needs of established IT operating units. A detailed exploration of the character, roles, and responsibilities of the PMO

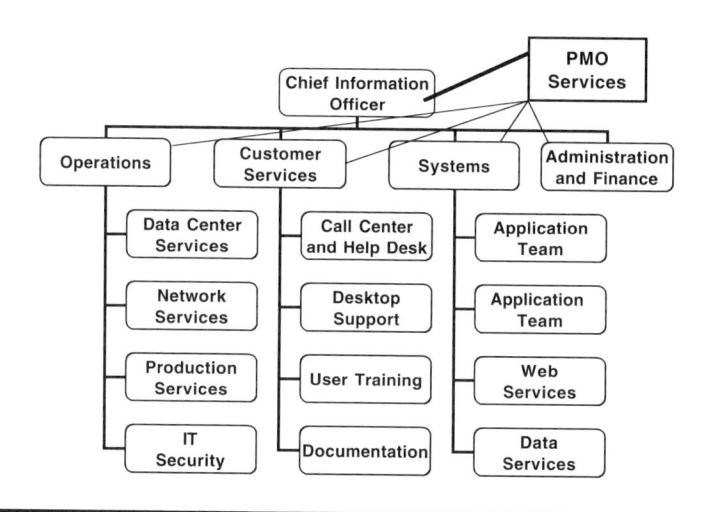

```
                                                    PMO
                        Chief Information         Services
                            Officer

    Operations      Customer          Systems       Administration
                    Services                        and Finance

            Data Center      Call Center       Application
            Services         and Help Desk     Team

            Network          Desktop           Application
            Services         Support           Team

            Production       User Training     Web
            Services                           Services

            IT               Documentation     Data
            Security                           Services
```

Exhibit 2 An IT Organizational Design Positioning the PMO

follows. But first, here are a few additional observations about the PMO's positioning within the greater IT organization.

No one in the information technology organization will contemplate the PMO, as described here, with indifference. Some will immediately recognize the potential benefits afforded by its creation, but most will remain suspicious or hostile. From the outset, it is essential to position the PMO as a support organization with the rest of the IT organization as its customer base. The PMO team must exemplify the operating principles and best practices espoused by IT management. More importantly, the team must behave even-handedly in its interactions with project and service delivery teams, appearing consistent in its adherence to operating unit standards while behaving flexibly in light of particular business or technology circumstances. To be sure, the team must sell its ideas and practices to encourage adherence and to avoid policing and blaming. In all instances, the PMO must take the high road in its own use of agreed-upon procedures, setting an exemplary standard for the greater IT team without lording over or overloading that team.*

At a more substantive level, the PMO should offer support services that might not otherwise be available within the IT organization, such as

---

* I must confess that as a PMO director, I have been guilty in the past of overloading my IT colleagues with new processes, procedures, and best practices, many more than they could possibly absorb, digest, and apply at any one time. A word to the wise: focus on those areas that need the most attention and work in small increments, moving from one successful application of process change to the next. Do not assume that your audience has the same capacity for or interest in process change that you may have.

the administration of planning and budgeting processes, project management and business analysis, benchmarking and performance measurement, customer relationship management, and operational reporting. By offering services that have recognized value but are somewhat neglected by the IT unit today, the PMO may win friends without treading on the territory of other IT departments.

Let the IT management team as a whole define and charter the PMO's scope of activities. This approach will provide the PMO from the outset with a sense of legitimacy and broad-based support. If possible, the office should report to the chief operating or executive officer within the IT organization accountable for customer delivery, and thus, one hopes, without affiliation to any particular IT operating unit. In this manner, the PMO may retain its objective position within IT while carrying the endorsement of the organization's executive leadership. Even so, for the PMO to succeed, it must have the recognition and acknowledged support of IT's rank and file. For this to occur, the office will need to demonstrate valuable contributions to the day-to-day operations of the greater team. Positive results will speak much more loudly than endorsements and policy statements.

Large IT organizations may already have a number of people in PMO-like roles and need only bring them together to create a center of excellence in the disciplines cited. The next two sections of this chapter will itemize the competencies required to staff a PMO in this fashion. However, many smaller IT groups will not have the luxury of a dedicated PMO staff. These teams will need to operate a more virtual PMO, in which individual members of staff, including the CIO himself or herself, take on the roles and responsibilities of the PMO without incurring the added expense of specialized personnel. In fact, IT organizations that have successfully deployed PMOs combine both operational constructs in their business model. On the one hand, they dedicate some of the day-to-day aspects of PMO functionality to a dedicated band of workers, while taking on more ubiquitous activities, such as customer relationship management, reporting, and performance measurement, as general management responsibilities. These they share across the IT management team. What is important here is not how the PMO is structured within IT, but that its service functions are available to the entire organization and that it is recognized as an internal service organization dedicated to raising the quality of the IT organization's service and project delivery.

The remainder of this chapter will explore some of the key roles and responsibilities in IT service and project delivery as these relate to the PMO model. Within this framework, the author will identify where the PMO might best add value in support of the IT organization. I will also provide a more detailed description of PMO roles and a functional business model for PMO operations. Lastly, I will consider the return on investment (ROI) to the information technology organization in establishing and operating a PMO.

IT SERVICE AND PROJECT DELIVERY ROLES

The focus of all IT activities revolves around service and project delivery.[*] In modeling the associated business processes, it is useful to think of the constituent parts of service and project delivery as a web of mutual commitments among sponsors, working clients, customers, IT delivery teams, and the PMO. As presented in Exhibit 3, these relationships interconnect, and through their transactions the various players in this scenario ultimately deliver value to the enterprise and its customers.

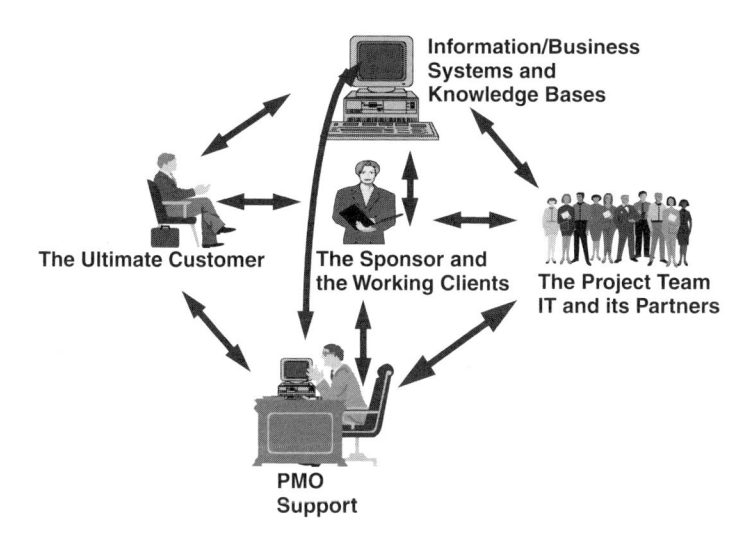

Exhibit 3   The Web of Commitments in Service and Project Delivery

---

[*] This chapter considers only a limited view of the roles and responsibilities of the total IT organization, those concerning the management and support processes for IT service and project delivery. For two more comprehensive models of IT roles, I offer the frameworks that I developed for the New England Financial and Northeastern University IT organizations respectively. I completed the former with the help of Bob Winn and the latter with the help of Denise Siciliano. See *The Hands-On Project Office,* http://www.crcpress.com/e_products/downloads/download.asp?cat_no=AU1991, chpt2~2~IT Competencies~model 1 and chpt2~3~IT Competencies~model 2. These models describe in considerable detail the skills required for effective service and project delivery within an IT organization. The second model considers how best to grow these competencies within IT, including the role that the PMO might play in such efforts.

Successful delivery depends on a shared sense of responsibility for and commitment to the desired outcome as understood by all parties within this web of relationships. The results of the IT service or project must be defined up front and subsequently assessed for compliance when actually delivered.

Within this framework, no service or project deliverable can exist without the direct participation of an active sponsor. A sponsor is the line-of-business leader who champions a particular IT service or product offering. A sponsor's support is essential to success in that he or she sells the value of a particular IT investment to the greater enterprise community and may formally fund the undertaking. Typically, this role is assumed by a member of executive-level management and the owner of the line-of-business or customer relationships directly impacted by the specific IT service or project under consideration. Sponsors provide a strategic enterprise perspective and an understanding of the overall scope of the business effort required for delivery. They are also well positioned to build bridges between key line-of-business stakeholders and the assigned IT team leaders (i.e., customer relationship executives, service managers, project directors, and project managers), to ensure that the proper resources are at hand to enable delivery and that service or project outcomes properly align with overall enterprise priorities.

Beyond providing focus, leadership, funding, and political support to the IT team, the sponsor also ensures that appropriate line-of-business resources, including working clients, are added to the project team. Furthermore, major IT undertakings often entail the rethinking of associated business processes. The sponsor is best positioned to recognize this need and marshal the political and organizational resources required to see the process changes to fruition. Many times, the sponsor will also conduct project reviews, sign off on project deliverables, and even preside over lessons learned at the conclusion of a service or project delivery cycle.

The sponsor, however, may have neither the time nor the inclination to participate more directly in IT-related work. He or she will typically delegate more direct responsibility for the success of the effort to subordinates in the impacted line of business. These day-to-day representatives of the business sponsor are the so-called working clients, providing detailed operational and tactical knowledge of the business activities in relation to the planned deployment of new information technologies. Unlike the sponsor, who may be totally uninvolved in project execution, the working clients will be intimately involved. As such, the role of the working client is multifaceted. He or she serves as a liaison between the business unit and IT, identifying and making available to the team business- and customer-specific expertise. The working client may also actively engage in project management, reviewing progress against the plan and helping to resolve operational and resource issues. Perhaps one of the key roles of the working client is to review interim project results and ensure that the emergent product or service aligns with the expectations of the sponsor and the needs of the business. In the author's experience, the availability and regular participation of working clients in any IT endeavor will make or break the project. Both the sponsor and the working clients should be on board before initiating any new undertaking.

Given the conflicting pressures that consume the lives of our colleagues on the line-of-business side of the enterprise, it is essential that IT maintain open and honest lines of communication with service or project sponsors and working clients. This must happen at various levels. Since a sponsor may have any number of projects under way with IT, someone must look after the entire portfolio of investments between that line-of-business sponsor and the IT organization. The sponsor is interested in one-stop shopping, that is, in having someone who can reach out across IT to answer questions, address problems, and advocate for the needs of the sponsor's business unit. Similarly, the sponsor and working clients benefit greatly from someone who understands the particulars of their business and can advise them on how best to invest their limited IT dollars and help them set realistic service level and project delivery targets.

At a more granular level, the working clients must interact with the IT manager who oversees the particular service or project of interest to that business unit. Here, the requirement is for detailed knowledge of the specific technologies and processes applied to achieve successful delivery. The working clients will expect an in-depth understanding of their unit's business processes, customers, and associated enabling technologies. They will also expect an IT counterpart who knows and understands the particulars of the project under way and has a general knowledge of all the information technologies that will contribute to delivery. Two very different roles within IT emerge from these requirements. At the portfolio level, the IT organization should establish a group of customer relationship executives (CREs) to address the need; at the specific project level, the solution calls for the creation of a project director's role.

The best CRE is someone from within the IT organization who is senior enough to recognize and appreciate opportunities for the information technology enablement of the business. The CRE need not be a technical expert, although in some instances, such as work that calls for highly integrated manufacturing systems or Web services, knowledge of the particular technologies in play may be essential. CREs should have strong people skills, especially listening, negotiating, and communication skills. When problems or changes in plan occur, CREs must be able to stand up to senior IT and business management, conveying the realities of the situation and its implications for timely, full delivery. At the same time, the CRE must know the entire IT organization and where to turn for help and advice. Typically, the CRE role calls for the maturity of judgment that comes with many years of service, within IT or the business itself.* Few IT organizations are large enough to field an independent team of CREs. Usually, the CIO and members of his or her executive team will each take on an assigned portfolio of line-of-business products and services, and serve as the IT organization's CREs.

Once CREs receive their customer assignments, their tasks are threefold. First, they will meet with their sponsors and working clients to collect service and project needs, always shaping customer expectations about IT capabilities and organization capacity. Second, they will keep their customers apprised of IT service and project delivery performance — both the good news and the bad. Third, they will bring opportunities for the further IT enablement of business operations to the attention of line-of-business executives. In short, the role of the CRE is to nurture the relationship between key customers and the IT organization through reporting, communicating, educating, and setting expectations. The particular tasks of the CRE might include the following:

- Collect and present the history of past services and projects, including success measures and related performance data
- Create and review with the customer IT's service level agreement (SLA) for the coming year
- Finalize agreement and signoff on the SLA and thereafter conduct periodic (e.g., monthly) reviews of actual IT performance as defined and measured by the metrics set down in the SLA
- Conduct monthly sponsor and working client working sessions to achieve the following:
  - Manage customer expectations
  - Gather customer feedback
  - Share end-user (i.e., ultimate customer) feedback
  - Explore emerging customer-generated requirements and IT opportunities
- Track and report on the overall status of all IT projects in the sponsor's IT portfolio

---

* When the IT team's relations with a particular line of business are frayed or when successful IT deployments call for a truly in-depth understanding of the business in question, the reader might consider recruiting someone with the appropriate skills and knowledge from within the business itself for the CRE role. Alternatively, you might find a now-retired working client who would find the CRE role interesting and not particularly taxing.

- Review, where appropriate, project scorecards* with the sponsor and the working clients, noting, communicating, and managing the following as need be:
  - Issues
  - Change orders
  - Emerging opportunities
  - Customer satisfaction feedback
- Bring new requirements and emerging opportunities, once these have been validated by the customer, to IT executive management for prioritization and approval
- Expand the scope of the IT portfolio for his or her client's line of business as appropriate
- Conduct satisfaction surveys at the close of a project with his or her sponsor and working clients
- Provide input to the IT planning and budgeting processes regarding the needs and expectation of his or her customers

By its very nature, the CRE role is very "top of the trees." It does not serve the more detailed management requirements associated with day-to-day delivery. Instead, it helps set the overall tone and direction of the relationship between IT and a key customer. A good CRE has a clear sense of what the line of business needs and what IT can do to help. The CRE's essential job is to manage expectations and maintain high-quality communication among IT service and product stakeholders. To that end, each CRE will work with any number of IT service delivery managers and project directors, who are in turn responsible within the IT organization for service and project delivery.

IT service delivery managers operate and maintain the hardware- and software-enabled services on behalf of line-of-business customers and the enterprise community as a whole. On the infrastructure side of things, they may maintain data centers, storage arrays, networks, printers, desktop systems, and the like. On the software side, they may develop, install, integrate, support, and enhance services, such as Web portals, voice response systems, and e-mail systems, as well as such business applications as general ledger and accounting, manufacturing, logistics and distribution, marketing, sales, and so forth. There may be dozens, even hundreds, of service managers in your IT organization. The CRE could not possibly interact with each of them individually.

Instead, each service delivery manager will establish enterprisewide service level descriptions and metrics for those who consume that service.

---

* The project scorecard is a snapshot of project status. Chapter 5 considers the scorecard in greater detail.

Through the PMO or perhaps the CRE himself or herself, the IT organization will incorporate this information into a single SLA for the review and signoff of business sponsors and working clients. Once agreed upon, the SLA will serve as the yardstick for assessing IT service delivery. (See Chapter 4 for details.) When the needs of the business change or when external or internal forces dictate an adjustment to SLAs, this information is communicated through the CRE, who renegotiates any changes to the SLA with the appropriate IT service delivery manager.

Because project initiation usually signals a major change to the use of IT and the underlying business process, information technology projects require a different sort of management process. Here, the CRE may report on the status of work for the benefit of the sponsor, but most of the action occurs between the working clients on the business side of the house and the project director and project manager on behalf of the IT organization. In this context, the IT project director is responsible for overall project delivery. As a practical matter, the project director is usually the IT service manager who will end up owning the ongoing maintenance and support of the new IT solution once it is delivered.* He or she has the authority to make decisions that will keep the project moving, works with the project manager in administering project plan tasks and resources, and keeps customers and upper management informed of project statuses. Most importantly, the project director ensures that his or her team is staffed with the right people to get the job done and that all of those people perform at peak efficiency and effectiveness.

After negotiating resource levels with the project sponsor and the project's internal and external partner providers, the project director works in concert with the project manager on commitments, plans, reports, metrics and all other planning and control tools. He or she participates in and at times leads customer meetings and briefings, keeps project working clients and perhaps sponsors apprised of current statuses, works with and oversees external service provider activities, monitors project-related contracts, maintains project budgets, communicates project issues to stakeholders, and chairs project management team meetings. When disagreements emerge within the project team, the project director should resolve such conflicts and keep the project moving. To accomplish these duties, the project director requires a general understanding of project management processes and techniques, an understanding of the end

---

* This is a practice that I recommend highly. If the project director is ultimately to own the fruit of his or her labor, this person will ensure that the product is well built and easily maintained. For that matter, the project director with an operational stake in the product's outcome will pour his or her experience into the solution's design. As a result, the IT product will tend to fit more appropriately the business process it is meant to serve.

customer and business processes, and an intimate understanding of the role that IT enablement plays in the particulars of the project.

For the project director, this is a long to-do list, especially when you realize that the person in question also typically runs a major IT service. Therefore, the project director will draw on the services of two PMO supporting resources: the project manager and the business analyst. The project manager provides continuous oversight of the project by communicating directly with the project director, the project sponsor, the working clients, and the IT executive team. He or she brings leadership skills and process expertise to the project, guiding and evaluating the performance of the project team. The project manager is responsible for the day-to-day activities of the project with regard to timely delivery, budget, and quality (i.e., meeting or exceeding customer requirements). At any given time, the project manager may be responsible for one or more projects. For small projects in which only a few people are involved, the project manager might serve as both the team leader and its business analyst.

The project manager brings the value of his or her particular discipline to each engagement, but this person also provides an objective perspective on how things are going within the project, with an eye toward how this particular project may impact or be impacted by other IT projects. This perspective is invaluable. All too often, those engaged do not have the leisure or vantage point to see the connections between their work and other activities across the IT organization. Their focus is on the particular technologies and business processes that serve as the focus for their project. Since the project manager's primary responsibilities concern process, he or she is better positioned to consider the implications of what is happening and how the actions of a particular project team might have downstream or upstream repercussions for some other IT effort.

The specific tasks of the project manager may include the following:

- Develop scoping statements and the definition of IT deliverables (not to be confused with the desired business outcomes that are defined by the business through the CRE or project director)
- Create work plans and budget estimates under the general direction of the project director
- Work with the project director in selecting staff resources and partner providers, and help to define the roles and responsibilities of individual team members
- Work with the project director on timetables, metrics, problem escalation procedures, and the like
- Conduct business process analysis and collect and document user requirements

- Manage day-to-day project work, including the following:
  - Monitor project plan compliance
  - Report on progress and issues against the plan
  - Work with external partner providers to deliver on their commitments
- Track issues and resource consumption for the project director
- Assist in the preparation of reports for IT management, project sponsors, and working clients
- Conduct project meetings and presentations
- Monitor the turnover of project deliverables
- Coordinate the SLA turnover
- Conduct lessons-learned sessions and customer satisfaction surveys
- Ensure that all project artifacts are posted to the appropriate IT knowledge base or archive

The project manager's toolkit ideally includes expertise in project and resource planning, vendor management, people management, and process management; specific business-related knowledge;* business process analysis; interpersonal communication; people skills; listening; conceptual modeling; and documentation and writing.

The list of project manager roles and responsibilities is long. The project manager may share this burden with one or more business analysts. As a member of the service or project delivery team, the business analyst possesses the skills to clearly identify business and customer requirements and to communicate and document these requirements on behalf of the team. The project team and its working clients then use this information to define or redefine organizational processes and IT functional system requirements. In executing his or her assignments, the business analyst will focus on the overall efficiency of the business process and its enabling technologies. The objective is to identify opportunities to improve business performance and to reduce the total cost of information technology ownership. To this end, the business analyst will typically do the following:

---

* Because it is useful, if not essential, for the project manager to develop of body of knowledge around a particular line of business or a family of information technologies, I have often assigned project managers their project portfolios aligned along one of these axes. Over time, the project manager develops a solid understanding of the business needs and issues common to his or her portfolio of assignments. The down side of the portfolio approach is that the project manager may get co-opted by the very line of business he or she services, thus losing objectivity. If this occurs, I would reassign the project manager to some other set of tasks.

- Document existing business processes and customer uses of IT
- Develop and document process flows in terms of a particular technology-enabled solutions (i.e., functional specifications)
- Develop project business and functional specifications and assist in the drafting of technical specifications
- Work with the project manager to develop statements of work, project schedules, and deliverable descriptions, and to issue tracking documents and management reports
- Prepare test scripts and quality assurance scenarios for the testing and release management processes

To do his or her job, the business analyst requires competencies in business process analysis, interpersonal communication, listening, conceptual modeling, documentation and business writing, project management, and the role of IT enablement in process design and engineering.

In summary, successful IT service and project delivery requires broad-based participation. The business side of the house provides the sponsors and working clients, who articulate the need for enabling information technologies and who work with IT to convert these requirements into IT services. IT CREs are usually senior managers who serve as overall service and project portfolio managers, working in close partnership with their line-of-business counterparts to shape the scope of IT assignments, set operational and tactical priorities, and allocate resources. The typical IT project director should be the same IT line manager who will maintain the new or enhanced product or service once it moves into production. The project director's tasks are supported through the efforts of a project manager and one or more business analysts,[*] who may or may not serve as part of the IT organization's PMO. Sponsors, working clients, CREs, and project directors may have occasional involvement in project delivery, but project managers and business analysts spend nearly all their time doing this work. In the author's view, these circumstances justify an IT center of excellence devoted to these management disciplines. The remainder of this chapter will consider, in greater detail, the function of and value proposition for such a center of excellence: the PMO.

---

[*] Another version of these roles and responsibilities, authored by Pat Erickson, manager of the Project Management Office, Information Services Division, Northeastern University, may be found on *The Hands-On Project Office*, http://www.crc-press.com/e_products/downloads/download.asp?cat_no=AU1991, chpt2~4~PM roles and responsibilities~example.

THE ROLE OF THE PROJECT MANAGEMENT OFFICE: MEASURING ITS ROI

So far, the author has described operating assumptions about IT planning, resource management, and service and project delivery. These models and examples may resonate with the reader's own experiences. Even so, a concern may arise that these models and processes carry too high a price tag, putting them beyond the means of smaller enterprises and their IT organizations. Admittedly, these methods carry their own level of IT commitment and investment. As I will demonstrate, however, the cost of developing and maintaining these processes prove insignificant when balanced against their payback: high-quality customer relationships and consistent success in the delivery of IT products and services. Even if one were to set aside these lofty goals as unobtainable, one might be satisfied at an operational level with an IT organization that can achieve the level of self-management prescribed here. Could the synchronization of service and project delivery management modeled here work within your organization? Do you and your team have the interest, commitment, and will to make it happen? These are valid questions that need answering. The mechanism of the PMO (virtual or staffed) and its associated tools and techniques can contribute substantially to enhancing your IT team's performance.

To assist you in selecting a course of action, the author offers a series of ROI measures of his own, posed as questions. My premise is that answering these questions in the particular context of your own business will justify your investment in a PMO. Since there are many ways to approach and scope the deliverables of a PMO, I will then consider the potential components and service offerings of a PMO in more detail before returning to the calculation of an ROI for the PMO. To begin, consider the following questions as indicators of the current state of your IT team's service and project delivery performance. What are the cost implications of noncompliance with a PMO model? Is absence of planning and poor execution depleting your human and financial capital and wearing your welcome thin with your employer? You judge:

- What is the state of your current IT planning process? Is the allocation of IT organization resources properly aligned with enterprise priorities? Are your executive customers concerned with the nature, direction, and growth of IT expenditures?
- If you have an IT planning process in place, what percentage of IT management time is devoted to that process? Are the leadership's time and effort in this process well spent or should their work be supplemented by a staff function?

- How are IT's relations with your key customers (i.e., sponsors)? Does the IT organization fully understand their needs and concerns? Do your sponsors in turn fully appreciate the constraints under which the IT team operates? What is the cost to IT (financially, operationally, and politically) in maintaining these relationships or failing to do so?
- How does your organization measure and report on its performance to those who fund your activities? Are your sponsors satisfied with both the metrics and your reporting process? How might a process that more accurately reflects the concerns and interests of your customers and is delivered in a more timely and effective manner affect your standing among sponsors and the enterprise as a whole?
- Are your customers satisfied with your service delivery? How do you know? Are your measures and responses reactive (i.e., after a service has broken down) or proactive (i.e., anticipating and correcting points of failure before they occur)? Are you managing customer expectations, and if not, what does it cost you every time you disappoint a sponsor? How many resources do you currently devote to reactive problem correction (e.g., call center, CREs, maintenance and support personnel)?
- Are your customers satisfied with your project delivery? Do you deliver your projects on time, within budget, and in keeping with customer requirements? How often are IT resources misdirected because of a misunderstanding about customer specifications? What are the true costs and impacts of project change orders? How do you know, and what are your measures of success in this area?
- How much IT staff time is devoted to finding out who on the team owns particular expertise, manages particular IT assets, or holds the responsibility for a particular area of service or project delivery?
- How much does your team pay for scrap and rework, for reinventing the wheel, and for failing to learn from past service and project delivery mistakes?
- What skills and experiences are needed for your team's success? How are these requirements captured and addressed today? How are they learned on the job? What internal resources do you draw upon to train and develop your staff — not just in technical areas but also in terms of their people skills?
- Does your enterprise suffer from the proliferation of information technologies, or do you promote architected IT solutions built around recognized standards? What is the total cost of IT solution ownership? Do you know? Do your sponsors know? Will the leveraging of core technologies, rather than the addition of new technologies, make a difference to you, your team, and the bottom line of your parent organization?

These questions explore the territory where a PMO or some other process-focused support team can make a difference for the IT organization. For example, the first two sets of questions consider technology planning, asking how these functions are conducted within your enterprise and whether the right roles and skills are in place to ensure the desired outcome. Next, the diagnostic examines the quality of your team's customer relationship management efforts, the benefits of attending to these activities, and the costs of their neglect. Next are considered the dimensions of successful service and project delivery, including coordinated management, reporting, and accountability. Your IT organization undoubtedly invests heavily in these areas, but are you satisfied with your return on the investment? Are getting the results and value for each dollar spent? More importantly, are your customers satisfied with the outcomes of these efforts? The diagnostic concludes by touching upon activities that relate to your team's communication, leveraging, and reuse of lessons learned. Effective and efficient IT organizations also invest in internally generated knowledge management, the fostering of team work, technology standardization, and the deployment of architected solutions. Together, these characteristics define a truly outstanding IT organization.

Unfortunately, the coordination of planning, customer relationship management, service and project delivery, performance measurement, and knowledge management does not readily fit within the skill set, focus, or perhaps the interest of most IT operating units. These tasks call for nonlinear-thinking, multitasking, process-oriented professionals with strong people skills. Product and service delivery would also benefit from the objectivity and perspective that transcends the siloed orientation of the rest of the IT organization. In short, they call for an independent body of experts whose mission is to support, enable, and facilitate the effectiveness of service and project delivery teams. This working group may serve as a stand-alone organization or may comprise contributors drawn from various IT departments and brought together as dictated by circumstances. Whatever your strategy for staffing this team, it is essential that the PMO have a clear charter and mandate and that the rest of IT understands and supports the rationale for bringing the PMO function into existence.

My own model for the project management office begins with a broad definition of roles and responsibilities, as depicted in Exhibit 4. The main roles of the PMO include

- Strategic responsibilities, i.e., coordination of such services as planning, benchmarking, management training and development, and other staff support functions
- Knowledge management of information specific to the enablement of IT delivery
- Service and project management

The Project Management Office Business Model ■ 45

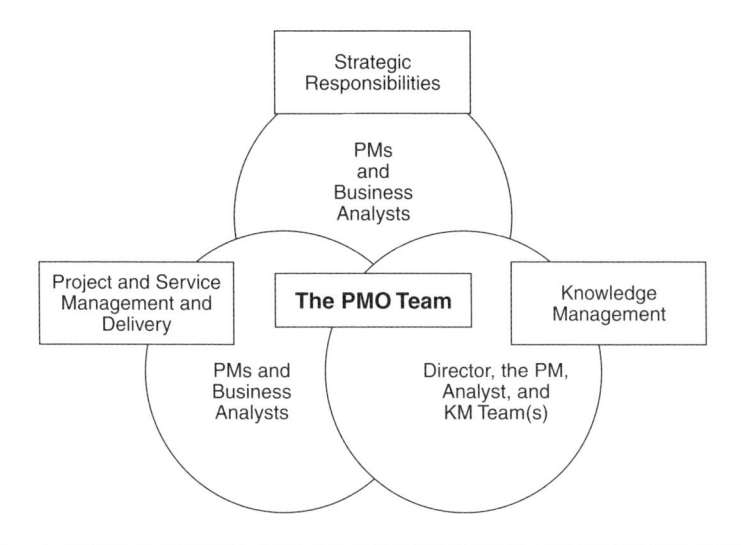

Exhibit 4 Areas of Competence and the Overlap of Responsibilities within the PMO

See Exhibit 4 for an illustration.[*]

Note that these domains deliberately overlap. Planning, measurement, and training inform and are guided by the results of service and project delivery. The experiences in service and project execution contribute to the vast body of tacit and explicit knowledge that will enable future IT team performance. The measures of past performance, customer requirements, and IT organization capabilities are all elements that feed the planning for future prioritization of IT customer commitments and resource allocations.

Each set of PMO activities involves different combinations of participants.[**] The strategic services function may be staffed by a single individual, possibly the manager of the PMO working with the leadership from across the IT organization. Based on the action plans generated through the planning process, the manager of the PMO would work with his or her project manager or business analyst team to define and assign portfolios of work to individual contributors. Business analysts and project managers,

---

[*] For a more detailed rendering of this diagram, see *The Hands-On Project Office*, http://www.crcpress.com/e_products/downloads/download.asp?cat_no=AU1991, chpt2~5~PMO Operations~framework and model.

[**] For a set of PMO project management job descriptions, see *The Hands-On Project Office*, http://www.crcpress.com/e_products/downloads/download.asp?cat_no=AU1991, chpt2~7~pmo~job descriptions. For those roles primarily concerned with knowledge management, see *The Hands-On Project Office*, http://www.crcpress.com/e_products/downloads/download.asp?cat_no=AU1991, chpt2~8~km~job descriptions.

working alongside their IT colleagues to deliver products and services to IT's customers, will generate artifacts, such as project plans, business requirements, technical specifications, training materials, and the like. These artifacts document the experiences associated with a particular IT deliverable and therefore constitute the team's recorded knowledge of that activity. By passing on these materials to the knowledge management function within the PMO, the office can catalog this information for repurposing and reuse at some future date.[*]

Even this brief introduction into the structure of a PMO suggests how the roles and responsibilities of participants overlap and reinforce one another. In effect, the PMO has many masters depending upon the service. For strategic and staff services, the PMO's customers are the IT executive team; for project and service delivery, they are those cross-functional IT teams accountable for projects and services; for knowledge management, the customers are first and foremost the rest of the IT organization, but perhaps also the enterprise's extended management team. See Exhibit 5.[**]

As Exhibit 5 illustrates, the PMO can and should touch all aspects of IT operations while remaining unattached to any particular operating unit. In its support role, the PMO may remain invisible to outside customers and partner providers. To maintain its independence and its comprehensive yet objective view of IT delivery, the PMO should report directly to the chief executive or operations officer of the IT organization. The specific tasks assigned to the office might include these:

- Ensure alignment between IT commitments and the enterprise's business objectives
- Collect, codify, and disseminate best practices among IT's service delivery and project teams
- Collect, document, and disseminate reusable components (such as project plans and budgets, commitment documents, technical specification templates, scripts and software components, and the like) to project teams
- Oversee the reporting and performance measurement needs of IT

---

[*] Chapter 7 and Chapter 8 discuss the knowledge management function of the PMO in detail and illustrate practical ways a modest investment in KM pays a healthy return to the IT organization.

[**] An electronic version of this model may be found on *The Hands-On Project Office*, http://www.crcpress.com/e_products/downloads/download.asp?cat_no=AU1991, chpt2~5~PMO Operations~framework and model. A more detailed breakdown of PMO roles and responsibilities is provided in a companion document on *The Hands-On Project Office*, http://www.crcpress.com/e_products/downloads/download.asp?cat_no=AU1991,chpt2~6~PMO work alignment~model.

The Project Management Office Business Model ■ 47

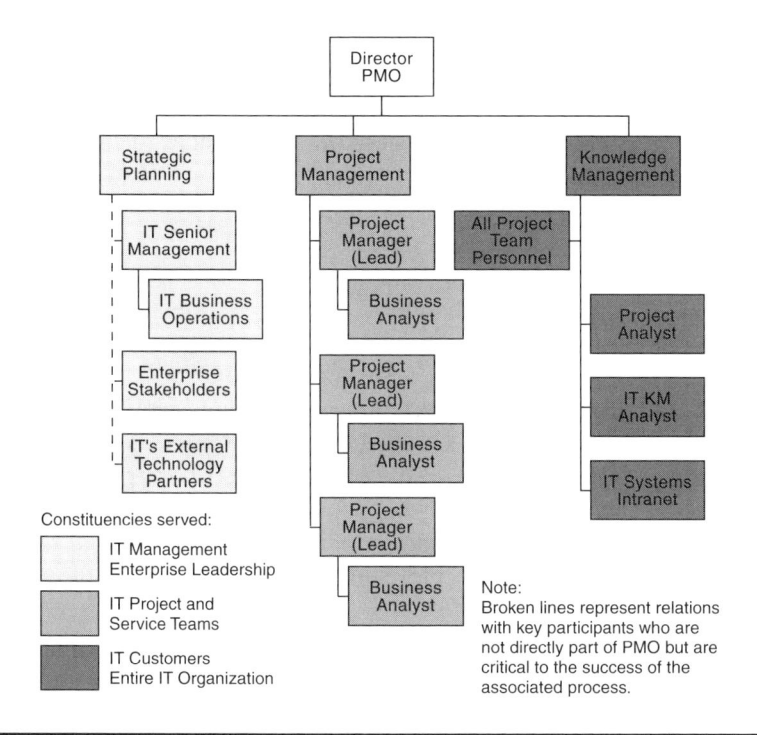

Exhibit 5   The Distribution of Labor within a PMO

- Oversee the articulation and updating of SLAs between IT operations and its various customers
- Support the CRE function
- Discover and share benchmarking and best practices data as this informs measures and models of IT service and project delivery
- Support IT management's continuous improvement and process change efforts
- Assist in the management development and training of IT personnel[*]

These assignments embrace the overall support of service and project delivery management, as well as the associated knowledge management component, leaving to IT operating units the responsibility for actual service and project delivery and for the underlying technical expertise to address customer needs. As suggested earlier and depending upon the resources allocated to the Project Management Office, the PMO function could also provide project managers and business analysts as support staff

---

[*] For a more detailed analysis of these activities, see *The Hands-On Project Office*, http://www.crcpress.com/e_products/downloads/download.asp?cat_no=AU1991, chpt2~3~IT Competencies~model 2.

to project teams. Project leadership would typically remain with the project directors, who would come from the information technology organization's line units and who would ultimately own the IT service emerging as a project outcome. As part of a project team, the services of PMO personnel might include the following:

- Assist in maintaining the project plan
- Maintain the project issue lists
- Maintain any project change of scope documentation (i.e., change orders) and the associated management processes
- Attend the weekly project team meetings and take minutes
- Attend the CRE and project director meetings with working clients and business sponsors as needed and take minutes
- Collect project artifacts (e.g., plans, scripts, best practices, system components) for reuse
- Promote reuse of IT project knowledge and encourage the use of best practices within project teams

Because the PMO focuses on process competence, leaving the technical side of the assignment to other IT personnel, the office's independence and objectivity position it to encourage the reuse of project artifacts and compliance with established best practices. Bear in mind that nothing about these processes is static. As the IT organization and its customers employ PMO templates and tools, they will adapt and refine these materials based upon practical field experience. PMO personnel will become the keepers and the chroniclers of this evolving institutional knowledge. Because they operate outside the reporting structure of other IT service delivery teams, PMO staff are in a position to advocate for and monitor the success of process improvements. In short, this team will help IT run like a business, facilitating and supporting the greater team's focus on successfully meeting customer requirements.

Although Chapter 7 and Chapter 8 discuss in some detail the role and value of the knowledge management function within the PMO, it makes sense now to consider how this function relates to the rest of the PMO. First, remember that in its support role, the PMO team exercises perspective into all aspects of IT operations. Yet, as a separate organizational unit, it is somewhat above the fray. PMO staff creates many of the artifacts that document IT organization processes but is sufficiently independent to recognize the less apparent dependencies among internal and external partner providers across the IT portfolio of projects. From this vantage point, the PMO will more easily identify opportunities for collaboration, leverage, and reuse, and it should be the first to recognize the implications of a particular project's slippage for other projects in the portfolio.

Furthermore, as a byproduct of their exposure to service and project delivery processes, PMO personnel will have direct experience of what works best in dealing with a particular customer, business problem, or confluence of information technologies. With this knowledge, the PMO team can develop process workflows, forms and templates, business rules, and even standardized technology components to better enable IT delivery. Over time, these activities can ensure the repeatable success of IT efforts. No one in a line position who is focusing on simply getting the job done would have the time or perhaps the interest to codify such knowledge into reusable forms and workflow diagrams. Codifying IT successes therefore becomes the domain of the PMO's knowledge management (KM) service.

To perform this function, the PMO's knowledge workers will require particular skills, such as an understanding of business process workflow rules, familiarity with knowledge management principles and associated technologies, and an ability to design form templates and to document business processes. In addition, these folks will need the tools common to the rest of the PMO team, namely solid interpersonal communication and people skills, listening skills, and oral and written communication skills. These are some of the particular tasks of the KM specialist:

- Create and maintain collaborative environments, like Lotus Notes TeamRooms, and document databases for projects teams
- Ensure that all vital project documentation is captured, retained, cataloged, and made accessible to those with a need to know
- Build and maintain an IT intranet site for the following:
  - IT organization planning content
  - IT standards and architecture content
  - Service and project management and delivery content
  - A resource allocation database
  - IT metrics and performance data
  - IT process best practices
  - Specific project content
  - IT asset inventories
- Manage access to and communications concerning IT project and knowledge databases
- Maintain forms and template libraries
- Maintain a best-practices library of case studies and benchmarking data
- Look for opportunities to bring knowledge artifacts and related PMO services to bear for the improvement of IT team performance
- Assist delivery teams to document results, processes, and lessons learned

Although most IT organizations get by without doing any of this, those in a position to rationalize processes and to leverage what they have learned from past performance will come out ahead of the game. In these ways, the PMO will help the IT management team to better execute overall planning and performance. They will support a variety of delivery team efforts, and they can provide an organization with valuable knowledge management services. The next section considers IT's return on the investment for establishing a PMO.

THE PMO VALUE PROPOSITION: AN INITIAL ROI ESTIMATE

The PMO's value proposition to the parent IT organization may be understood in different ways:

- Resource conservation — what does the PMO provide that might otherwise fall to more expensive and less experienced (in those disciplines that are the domain of the PMO) IT executives or overtaxed IT line managers?
- Risk management — what insurance do the services of PMO afford that might otherwise lead to service or project delivery failure, and hence to serious customer dissatisfaction with IT?
- Accountability and customer relationship management — what value will a more accountable IT organization and a higher level of customer understanding and confidence bring to the information technology organization's ability to satisfy its sponsors and therefore obtain the political, human, and financial commitments required from the business for successful IT delivery?
- Doing the right things — what is the cost to the enterprise of embarking on low-priority or simply wrong-headed capital IT projects?
- Doing things right — what is the cost of shoddy workmanship or scrap and rework within IT, especially when it leads to customer confusion (calling for greater call center and field or desktop support) and to more general and pernicious dissatisfaction with IT results?
- Total cost of information technology ownership — what is the value to the IT organization, and to the enterprise more generally, of streamlined technology platform choices and in a standardized IT approach to particular business problems and needs?
- Internal harmony, communication, and collaboration — what is the value to IT management of delivery teams that work together better, communicate more effectively, and employ timely and accurate information about IT operations?

Admittedly, it is not easy to quantify the ROI on any of these factors. On the other hand, most IT organizations face some, if not all, of the underlying issues reflected in these metrics. For example, the costs associated with scrap and rework, and the pursuit of improperly scoped IT projects could, if avoided, fund the type of PMO described in this chapter. Indeed, I would argue that by investing in a PMO, the reader will avoid much of the project life cycle's financial exposures while benefiting from less quantifiable quality improvements in delivery. Let us consider how all of this translates into a calculation of PMO value to the IT organization.

First, it bears repeating that there are three legs of the Project Management Office:

- Internal IT management and related executive team support services
- Service and project delivery support
- IT knowledge management services

See Exhibit 6.*

Assuming that your PMO takes on a range of assignments similar to those depicted in Exhibit 6, it is possible to model at some level of detail the benefits accrued through the PMO team versus the costs incurred in their employment. The net of benefits less costs will equal the value of the PMO. The accompanying illustrations provide the basis for modeling the potential value of a PMO. The author's worksheet includes the following analytical elements.** See Exhibit 7.

The far left column lists factors potentially impacted by PMO services, such as planning, service delivery management, and staff training and development. The second column captures IT organizational benefits — either cost avoidance or cost savings — incurred by IT thanks to the efforts of the PMO. For example, this column would include the estimated financial value of the time otherwise spent by IT executive management in preparing plans and status reports at the expense of tasks more in keeping with their roles and responsibilities, as well as the value of training delivered by the PMO to staff versus the cost of external partner provider delivery. Similarly, if the PMO saved a project delivery team time and money by repurposing knowledge gained from other projects or past IT activities, this calculated value might be added to the second column.

---

* For more details concerning the tasks and deliverables of the PMO, see also *The Hands-On Project Office,* http://www.crcpress.com/e_products/downloads/download.asp?cat_no=AU1991, chpt2~6~PMO work alignment~model.

** For an electronic version of the ROI worksheet, see *The Hands-On Project Office,* http://www.crcpress.com/e_products/downloads/download.asp?cat_no=AU1991, chpt2~9~PMO value calculation~model and template. For a hardcopy version of this tool see Appendix C.

52 ■ The Hands-On Project Office

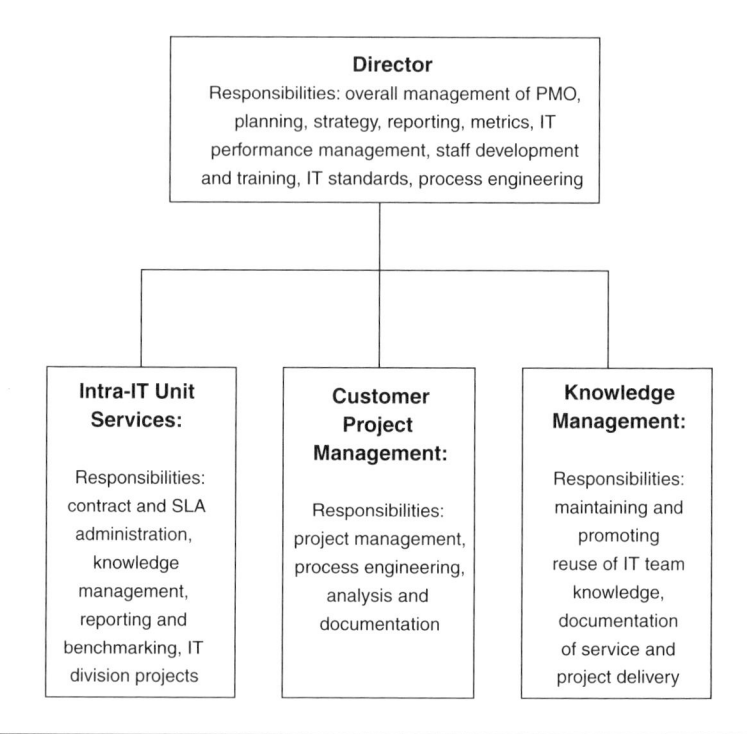

```
                    ┌─────────────────────────────────┐
                    │           Director              │
                    │ Responsibilities: overall       │
                    │ management of PMO,              │
                    │ planning, strategy, reporting,  │
                    │ metrics, IT performance         │
                    │ management, staff development   │
                    │ and training, IT standards,     │
                    │ process engineering             │
                    └─────────────────────────────────┘
```

Intra-IT Unit Services:	Customer Project Management:	Knowledge Management:
Responsibilities: contract and SLA administration, knowledge management, reporting and benchmarking, IT division projects	Responsibilities: project management, process engineering, analysis and documentation	Responsibilities: maintaining and promoting reuse of IT team knowledge, documentation of service and project delivery

Exhibit 6   The PMO Organization

The next two columns address the areas of service and project delivery risk. How might the effective performance of the PMO mitigate such risks? In the case of service delivery, for example, a thorough SLA process that is well communicated to the customer will help manage customer expectations. From this SLA process, customers should know what to expect from particular IT services, reducing call center traffic and confusion over support. This could translate into marginal cost savings for IT and better relations between IT and its customers. On the project side of the house, risks are more tangible, and therefore the benefits of mitigation are more

Exhibit 7   The PMO Value Calculation Model — Part 1

Value/Cost Categories	Amount of Financial Benefit/ Non-PMO IT Costs	IT Costs Avoidance Associated with Service Delivery Risk Mitigation	IT Costs Avoidance Associated with Project Delivery Risk Mitigation	PMO Investments in Services and Risk Mitigation	Outcomes (Net Value of Positive Outcomes and Risk Avoidance Less PMO Costs)	Comments

readily quantifiable. In valuating the relative benefit of PMO services, include the following risk measures in your model:

- Strategic risk (impact to business if application is not delivered, if it fails in production)
- Financial risk (soundness of budget, security of funding, actuals in line with budget)
- Project management risk (scope creep, cost and time overruns, people)
- Technology risk (performance, deployment, support, integration, interoperability, standards, expertise, competencies)
- Change management risk (change process documented and followed business requirements stable)
- Quality risk (requirements clear, project results traceable to requirements, reviews with customer in place)

While some of these factors are more difficult to quantify than others, all contribute to the cost of project outcomes. To the extent that the PMO's services mitigate these risks, the office has added value and avoided serious costs for the IT organization and the enterprise. The next column in the model is set aside for the inclusion of the actual PMO costs incurred in the delivery of particular service category component. By subtracting PMO costs from the prior three columns of benefits, one may derive the incremental value of the office.

Most of the opportunity costs and benefit categories outlined in the model are self-explanatory. See Exhibit 8.

The planning work avoided by IT management but conducted instead by the PMO has tangible value to the organization and its ability to align with the enterprise. Without these activities, the organization's business leaders may be unwilling to fund IT at a level commensurate with the needs and commitments of the organization to its customers. Previously, I have discussed the ways my model captures PMO effectiveness in support of IT service and project delivery. The template itself provides space for any number of SLAs or capital projects in this regard. To these I have added two other areas of quantifiable PMO contribution: technology platform and IT staffing cost savings or avoidance. Through the promotion of technology standards, reuse, and architected solutions, the PMO champions the rationalization of IT options within the enterprise. If adopted by IT and its customers, these standards will serve as a substantial, ongoing source of savings and cost avoidance, favorably impacting the total cost of technology ownership. Similarly, knowledge management will reduce the time and hence the costs of bringing new employees up to speed and help them to avoid the mistakes of their predecessors. A well trained and informed staff is happier, more productive, and less apt to turn over. This, too, will save the IT organization money and contribute to

Exhibit 8   The PMO Value Calculation Model — Part 2

Planning Process

Requirements
Benchmarking
Metrics and targets
Template prep
Process management
Presentations
Documents and updates
Reporting

Service Level Management

Customer Group A

Requirements
Metrics and targets
Template prep
Process management
Documents and updates
Reporting

Project Delivery Management

Project A

Strategic risk
Financial risk
PM risk
Technology risk
Change management risk
Quality risk

Technology Platform Standardization and Rationalization

Hardware costs
Software costs
Support costs
Contractor costs
Customer service costs
Staff training costs
Other costs (facilities, backup, etc.)

Exhibit 8 (continued) The PMO Value Calculation Model — Part 2

Staffing Costs

Management training
Technical training
Retention
Recruiting

more reliable performance. Obviously, many other factors enter into the effective delivery of these solutions to the IT organization, but the PMO team will be central to the solutions' development and implementation.

Admittedly, no one with any sense of balance sheet accounting will view this valuation model as a rigorous analytical tool. The underlying economics of the typical IT organization are too complex to capture in a single spreadsheet. Rather, my approach to an ROI calculation for the PMO serves as the basis for an informed discussion concerning the office's value to the information technology organization and the enterprise. IT service and project delivery is fraught with risks, and those on the firing line do not always have the time or the skills to address these risks. My model outlines the scope and nature of the PMO's impact in mitigating these dangers. As discussed, the tool provides a means to estimate contribution in any number of areas. Even though my model fails to satisfy the need for mathematical rigor, it should be clear that, even by conservative measures, the benefits of the PMO far outweigh its costs. The author's evaluation matrix will assist you in clarifying the true value of the PMO's contribution to your organization.

Later, this book will return to the subject of the ROI of the project management office. First, it examines in more detail the role that the PMO should play in IT team service and project delivery and knowledge management. My approach includes general narrative combined with applied case studies. The tools and examples offered in the accompanying Web site (http://www.crcpress.com/e_products/downloads/download.asp?cat_no=AU1991) will allow the reader to draw upon and adapt examples to fit them more appropriately into the context and operations of your own IT organization. Next, I will examine the role of the PMO in IT planning and resource alignment. Then I will proceed through chapters addressing service delivery, project management, and requirement gathering. Lastly, I will explore two PMO-maintained knowledge management applications before returning to the matter of the project management office's return on investment.

3

ALIGNMENT AND PLANNING — DOING THE RIGHT THINGS

INTRODUCTION

Information Week and other respected IT industry trade journals regularly publish lists of the best chief information officers (CIOs) and IT organizations. Admittedly, some of these competitions are mere beauty contests or have a contestant pool biased toward very large enterprises with enormous IT budgets. However, drilling below the surface of these industry surveys soon reveals at least one attribute that distinguishes the winners of these contests: the strength of the alignment between the information technology needs of these enterprises and what their IT units actually deliver. In other words, properly aligned IT organizations are seen as regularly delivering the right stuff to their customers.

In this sense, alignment is a tangible quality that may be measured in terms of the degree to which the IT organization takes direction from and delivers on its commitments to the business, contributing tangibly to the enterprise's bottom line and to customer satisfaction. At the same time, alignment is also about perception: business managers believe that they truly provide direction to the IT organization and IT leaders consciously determine their priorities and resource allocations in light of this input. To be sure, alignment along these lines is essential if the corporation's investment in IT is to enable the realization of business goals and objectives. Alignment does not just happen. The successful application of IT within the enterprise requires the coordinated and collaborative efforts of both business and IT management. When these parties are aligned, the ownership of information technology initiatives is shared, and the costs and risks associated with IT deployments are better appreciated and taken into account by all corporate stakeholders.

As important as alignment between business and IT is in any enterprise, achieving such an alignment does not come easily. First and foremost, the effort requires the full backing of the enterprise's leadership team. Management must recognize that getting it right requires an investment of its own time in defining current and anticipated information technology needs and then in prioritizing proposed initiatives to address those needs. Because available IT resources are invariably constrained, decision makers must limit the range and scope of the IT undertakings in a given fiscal year, and even jettison pet projects that do not make the cut.

For their part, IT managers must understand the business; intuit competitive opportunities and operational needs; and discuss the real benefits, costs, and risks of IT investment options in terms that the business side of the house can appreciate. They must also ensure that each chosen project encompasses the total cost of ownership (TCO) inherent in that undertaking. This means that they must be clear with their line-of-business colleagues about the added infrastructure enhancement and upgrade costs, IT staffing and support costs, and ongoing maintenance costs associated with any envisioned initiative.

Understandably, neither business nor IT leaders have the time, the interest, or perhaps the courage to master the details of their counterparts' areas of expertise. Somehow, this gap must be bridged and a shared vision of the role of IT within the enterprise established. By taking a proactive role in framing the alignment discussion within the enterprise, IT management can ensure that the business team focuses on realistic information technology opportunities and issues. IT leaders must also ensure that their business colleagues are well informed to make proper choices among IT investment alternatives. To this end, the IT team would be well advised to adapt the IT internal economy model discussed in Chapter 1 to frame this complex discussion. In addition, the leadership team must draw on the enterprise knowledge of its customer relationship executive (CRE) cadre and the process expertise and support of its project management office (PMO).

In employing the internal economy model, enterprise management is obliged to examine systematically the extent of current IT team resource commitments. After accounting for all nondiscretionary IT expenditures, including the incremental, ongoing costs of projects already in the pipeline, management can project what resources may be available for new initiatives. In brief, managers will know where the corporation currently sets the bar for IT investing. If the priority list that emerges from the alignment process exceeds this amount (which is almost always the case), the leadership must either cut back the list or add to the IT organization's resource base. Alternatively, management may agree to reduce the current investment in ongoing IT operations and even eliminate some offerings to enlarge the pool of funds available for new investments. These are often tough decisions. The advantage of a planning and alignment process is that final choices are made collectively and owned by the very lines of business that will fund the final list of new IT projects. Thus, the pain of denial or change associated with the adoption of technology initiatives for the year is shared by all stakeholders in the process.

This chapter explores a systematic approach to IT alignment and planning within the enterprise. Because it is very much in the IT organization's interest to get this process right, I advocate a proactive approach in which the IT leadership that engages the business side of the house in a constructive, informed dialog about enterprisewide IT investment options, and in which PMO personnel serve as working session facilitators, scribes, and timekeepers.

Beware! These conversations can be perilous because they could let loose unrealizable customer expectations. To avoid such circumstances, the process must educate participants about the fiscal, technical, and human resource constraints of the situation. At the same time, it must afford the business side of the house final say over initiative priorities. In this way, the ownership of process outcomes, as well as ultimate IT service and project deliverables, will be shared jointly by the enterprise's business and IT leaders. Finally, to ensure a common, clear understanding of the decisions made, the alignment and planning process must be documented to identify those accountable to articulate clearly the commitments themselves and the metrics of delivery.

Getting the business side of the house to participate in the process presents its own challenges. Even if enterprise management is committed to playing, managers may not have a deep enough understanding of information technology to differentiate among options. The next section of this chapter explores techniques for involving the nontechnical management team in defining IT goals and objectives. Then comes consideration of a number of approaches to achieving alignment, borrowing heavily on the author's own field work and on a methodology developed by Tom Murphy.*

Lastly, the chapter offers a model for the documentation of and reporting on IT planning and alignment commitments. This section will also speak to the need to market the IT plan, educating line-of-business management and other internal IT customers about the scope of these commitments, how results are to be measured, and what operating values

---

* For an overview of Tom Murphy's approach to IT portfolio planning within the enterprise, see "A Portfolio Planning Approach for IT Investment," *Enterprise Operations Management Journal*, 42-10-50 (August/September 2003).

and assumptions the IT organization intends to bring to its service and project delivery efforts. A series of simple frameworks will help focus and drive discussion and provide a comprehensive reporting tool for capturing and communicating IT plans and priorities across the enterprise. PMO personnel are ubiquitous in these activities but remain largely behind the scenes.

GETTING THE BUSINESS TO SET IT PRIORITIES

The key to any successful IT planning exercise is the careful framing of the discussion. At its core, this process must focus on the enterprise and not on technology. It must provide ample opportunity to explore possibilities but must also remain credible from the perspectives of business operations, finance, technology management, and project execution. Building castles in the air may excite the audience — only to provide frustration later, when visions of the future are scrutinized through the harsh lens of real-world fiscal and technical constraints. It therefore behooves IT leadership to script planning events with care, always framing them within a context that is familiar and comfortable to the organization's business leadership.

To this end, I like to begin any planning exercise with a set of questions that requires the planning team to focus on key issues and the right context for conversation.* This script might include such questions as the following:

- What is our mission as an enterprise?
- What are the barriers to that mission's realization?
- Who are our customers now and how will (or should) this change in relation to our mission? What are these customers looking to us for now? What will they look for in the future?
- What are the top goals and objectives of the enterprise as a whole? Of each key business unit? Of the board(s) of governance? Of the executive management team?

---

* The players in a particular enterprise's business and IT planning exercise vary from one organization to the next. In general, the team should include the chief executive officer or president; the chief financial officer; the responsible executives from each line of business; the chief information officer; and any other planning, operations, and IT personnel (e.g., IT CREs) with germane expertise or a significant stake in the discussion. In larger enterprises, these working sessions may occur between the executive leadership of a particular line of business and representatives of the IT organization. The director of the PMO, and perhaps members of his or her team, will provide facilitation and staff support for these activities.

- If these goals and objectives draw the enterprise in opposite directions, how can they be reconciled?
- What other barriers stand in the way of realizing these goals and objectives?
- What are our competitors doing? Should we be concerned?
- What other forces, such as the regulatory environment, are at play? How will changes in these forces affect our business and our plans?
- How might IT enable the realization of these goals and objectives? Assist in taking down barriers to the success of the enterprise? Afford new opportunities to the business?

These are open-ended questions meant to stimulate thought and discussion. To the extent possible, avoid an overt discussion of technology. Too often business colleagues will more willingly discuss the latest in IT trends and gadgets than deal with the real, tough, complex, and messy process and performance issues that confront their own operating units. This initial conversation needs to be exploratory, and though open to all possibilities, balanced by the realities of where the enterprise finds itself in its marketplace. To stimulate and facilitate these deliberations, a version of the Osborne/Kesner transformation diamond is helpful. See Exhibit 1.

The external environment - a.k.a. market forces

- Internal processes of business change and inertia
- Business process evolution
- The Enterprise
- IT innovation and change
- Existing information technology systems

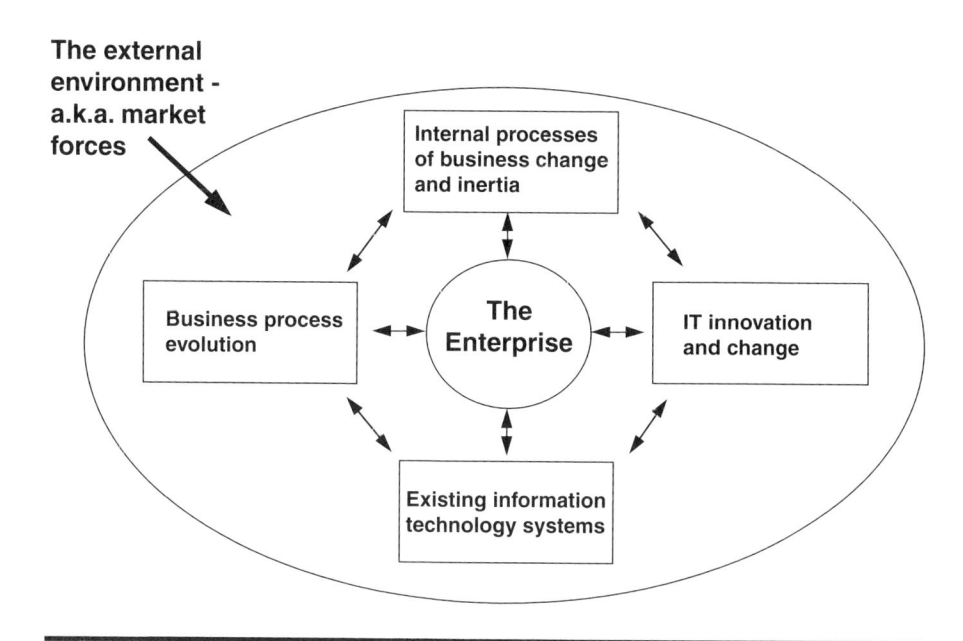

Exhibit 1 The Osborne/Kesner Enterprise Transformation Forces Diamond Revisited

In this version of the framework, some of the diamond's corners have changed. The enterprise sits in the middle of the drawing and is acted upon by four countervailing forces:

- Internal processes of business change and inertia
- IT innovation and change
- Existing IT systems
- Business process evolution

All of these activities (actions and reactions) float within an external business environment that in turns acts upon, shapes, and limits the forces at play within the enterprise. The transformation diamond helps discussants keep in mind the factors to be considered in identifying IT investment opportunities.

First and foremost, the planning team needs to understand the challenges faced by the enterprise and its constituent parts, hence the positioning of the enterprise at the center of the model and of the market place as a backdrop surrounding everything else. The aforementioned list of questions or something similar will draw out this context nicely. Next, ask what needs to change in the business to accommodate or address these dynamics. All too often, the business leaders assume that the introduction of a revolutionary technology will fix what is ailing or challenging the corporation. But technology can only enable sound business processes and practices. Thus, if the management team is looking to change, it must first grapple with the business process innovation corner of the framework.

Having established some sense of where the organization is headed with its process transformations, leadership must then ask itself these questions:

- What are the barriers within the various lines of business to such changes? For example, is the enterprise's culture adaptable?
- Do employees have the right skills?
- Are corporate rewards and recognitions aligned with the envisioned ways of doing business?

In a work environment of well entrenched business practices governed by labor contracts and the like, moving the enterprise to a new way of delivering goods and services may prove extremely difficult, regardless of the information technologies at hand. Taking the planning team through the discussion to this point exposes many of the critical challenges to be overcome in selecting and deploying business-enabling technologies. Note that the discussion has yet to touch upon the subject of information technology itself.

With all of this foundation building, IT leadership may now join the conversation with suggestions about how recent developments in information technology may assist in addressing the emerging needs of the enterprise. Here, timing is crucial. By ordering the conversation as described, even at this early stage in the planning process the IT team has aligned its view of technological innovation with enterprise business drivers, issues, and priorities. Be aware that your business colleagues will hear what they want to hear during these brainstorming discussions. If you imply that a particular technology may be available to address the need, they will understand you to say that it can be had immediately. Possible but highly challenging IT solutions may be received quite differently by those without technical experience. Avoid the hard sell and hype. Be realistic, identifying both the benefits and pitfalls of the various IT strategies under consideration.

Although this sort of exploration is vital to the development of aligned business and IT planning, your enterprise's management team may not feel well enough informed about the state of corporate IT investments to have such a conversation. Managers may need a broad overview of where things stand within their departments concerning IT products and services currently in place or on the way. Thus, to inform and stimulate the planning process, the IT team may need to do some additional spade work. A simple inventory tool can categorize the current state of IT investments in terms of sponsorship.* Most projects are easily categorized as enterprise, associated with a particular line of business, or infrastructure (i.e., IT-driven).

This inventory tool should be organized in a meaningful way to reflect the larger enterprise and the nature of IT spending. For example, if the lines of business drive IT projects, the left column of the inventory worksheet should list each line of business and then the primary services (i.e., business processes) operating within each unit. Thus, the finance division might include general accounting, accounts receivable, accounts payable, budgeting, fixed asset management, and so forth. Alternatively, where processes are highly redundant from business unit to business unit, you might organize your inventory by functions, like customer relationship management, product design, and manufacturing, listing the business units that employ those processes underneath. The inventory next captures the following information:

---

* See *The Hands-On Project Office* at http://www.crcpress.com/e_products/downloads/download.asp?cat_no=AU1991, chpt3~1~IT~inventory~example for an illustrated application of the tool. See chpt3~2~IT~inventory~template for a blank template that can be adapted to the needs of the reader's organization.

- Envisioned or proposed priority status of IT work to be done for this business unit or on behalf of this business process (with coding such as P=priority level; D=done; H=high, address immediately; M=medium, address in 2xxx; L=low, address in 2xxy). The list may include prior commitments and proposed commitments. Let your customers guide you in setting the dates initially; the management review process will firm them up for incorporation in IT plans.
- Business process is documented (yes/no) — essential information if the enterprise intends to replace legacy systems or in any other way alter the process in conjunction with the introduction of new information technologies.
- Business process has forms (yes/no) — these forms will be evidentiary to the process and may be replaced or superannuated by the introduction of IT systems.
- Current IT system(s), in place or planned as the case may be.
- IT system or service owner — one business unit may own the system while many units use it.
- Comment field for the status of work completed, under way, pending, or proposed. Color-coding entries of special interest makes them stand out during the management review process.

Given its comprehensive perspective of current and upcoming IT organization commitments, the PMO is well situated to prepare this document for IT management's use. The value of this exercise is multiple. First and foremost, it provides a current, top-of-the-trees view of all IT work under way or planned. It serves as a summary portfolio description of the IT organization's commitments. Second, with this picture in hand, management can more easily see where investment dollars are going. Such information naturally leads to a debate over alignment, balance, resource allocations, and enterprise priorities.

Third, it opens the door for a discussion of IT value in terms that relate nicely to our internal economy model. If the inventory makes clear that all immediate IT resources are committed, any further investments must come from a realignment of resources or from the augmentation of IT funding. Do not expect your business colleagues to open their wallets. More realistically, they will work with you to reduce current spending to reallocate resources to projects of higher priority.* Lastly, the inventory approach forces management to recognize, if it has not already done so, the centrality of information technology to its business and the role that

---

* *The Hands-On Project Office*, at http://www.crcpress.com/e_products/downloads/download.asp?cat_no=AU1991, includes a tally sheet for tracking the prioritization and reprioritization of IT projects. See chpt3~3~priority worksheet~template.

the enterprise's various lines of business must play in setting and supporting the direction of IT investing.

To round out this exploratory exchange of views, the management team should examine the implications of particular IT options to the enterprise. Just as there are established business practices to overcome if the organization is to change, new technology investments must be viewed within the context of existing enterprise IT capabilities. For example, if the option on the table requires the business to replace its network infrastructure, its server complex, or its legacy software code, what is the true TCO of such a decision? Does the IT team itself know enough to appreciate the implications (i.e., the ripple effects) of such a change? Are enterprise personnel sufficiently skilled to operate and maintain these new technologies? What are the human resource implications of the envisioned technological changes for the IT organization? Will IT need to retrain or restaff?

In other words, when looking at new IT investments, your organization must also consider its past investments and the embedded base of its IT infrastructure. Just as there are no business decisions in a vacuum, there are no neutral IT options. Each will have its pros and cons in terms of how it maps against the needs of the business and its fit within the existing complex of the organization's IT offerings and capabilities.

The transformation diamond affords a framework for drawing out these variables and demonstrating some of the more obvious connections between the workings of the enterprise at large and a business process change or an IT investment decision. However, the diamond does not in and of itself force the discussion to the level of detail required to determine the TCO of any particular IT investment option. To that end, I offer another simple tool to focus and discipline these more detailed discussions. If a particular IT undertaking clears the first hurdle and makes it to the wish list, turn next to this tool to gather the data required for the IT prioritization process. Here too, the PMO's project managers and business analysts can conduct the requisite information gathering and preparatory analysis in anticipation of a more detailed review process.

This data may be collected for analytical purposes in some vehicle like the Information Technology Project Request form.* This document facilitates the capture of the following pertinent information about any proposed IT venture:

- Project name
- Sponsor's name

---

* See *The Hands-On Project Office*, http://www.crcpress.com/e_products/downloads/download.asp?cat_no=AU1991, chpt3~4~IT Project Justification~template. For a hardcopy version of this tool see Appendix A.

- Names of all working clients
- Total funding request by fiscal year (to be derived from subsequent work sheets)
- Chronological list of envisioned project milestones
- Project cost table laid out by quarter, internal staffing head counts and costs, vendor costs, hardware and software costs, and any other project related expenses
- Another table that defines project roles and responsibilities and the head count associated with each project role
- Questions and a table for ROI and benefits calculations
- Strategic alignment checklist
- Risk management checklist

The request questionnaire asks tough questions and, when completed in full, provides all the fodder for something like the Murphy ranking process described later in this chapter. The rigor required in the request questionnaire will help thin the list of fanciful project ideas and proposals that offer only a marginal impact on the core business and the enterprise's strategic objectives. The form and its associated process can help take a lot of the politics out of priority discussions. If the value to the organization is below an agreed-upon ROI hurdle rate, even the sponsor of that idea should agree to its being dropped from further consideration in favor of more promising options.

However, the important contribution of this process is that it gets the issues out on the table, with their complexity, risks, and costs exposed for scrutiny. Certainly, any presentation of risks, costs, and ROI will be scrutinized by those with competing requests. The justification tool levels the playing field, requiring each sponsor to develop credible data in support of his or her proposal. If all goes well, your team will use this process, or something like it, to come up with a reasonable list of potential opportunities. The pool of ideas so generated will probably range from obvious and necessary investment decisions to large, optimistic ventures. The next step will be to work with management colleagues and the IT team to define and narrow this list of options. Ideally, you will end up with a tally of prioritized action items commensurate with available organizational resources and hence ready for the review and approval of your enterprise's governance process.[*]

---

[*] As part of an annualized planning and budgeting cycle, each organization will have its own approval process. Perhaps the same line-of-business partners engaged in this brainstorming exercise are in a position to approve and fund the final short list of projects. Alternatively, the list may have to go through some sort of executive review, even a board of directors' review. Before launching any planning and alignment discussions, be well versed in how these processes work within your own organization.

Bear in mind that what you have in hand is a wish list. The request questionnaire provides estimates and the rationale for individual investments — not prioritized choices from the investment pool as a whole. Pursuing the suggestions on that list without vetting them further will put your own team and the enterprise at risk. Once your colleagues understand more fully what it will take to achieve the deliverables that emerged from your brainstorming exercises, they will be in a position to appreciate the TCO of any particular choice and to make the necessary tradeoffs or scope revisions among the options.*

It is essential that the business leadership — not IT — makes any final choices. To that end, the leadership of the IT team must ensure that line-of-business colleagues have the information in hand to make informed decisions without dwelling too heavily on the more technical side of the question. In other words, to proceed with any sort of reasonable prioritization process, management requires a clear understanding of the business opportunity afforded by each technology investment on the wish list, as well as the associated risks of proceeding with that undertaking. The next section of this chapter explores how to position the discussion for priority, how to build consensus around a short list, and how to ensure that your line-of-business colleagues buy into the final decision.

GETTING TO "YES" IN SETTING IT PRIORITIES: AN APPROACH TO BUSINESS AND IT ALIGNMENT

Even in top-down, hierarchical organizations, decisions require buy-in to be successfully implemented. Therefore, any major IT investment impacting the entire enterprise needs the endorsement, if not the championing, of the greater organization to succeed. Indeed, without the backing and direct participation of the business side of the house, few IT projects will ever realize their goals. For that matter, most significant IT projects entail business and technical requirements analysis, process reengineering, system and service deployment, and end-user training and support. These activities call for the direct involvement of line-of-business personnel. If the business leadership is placed in the situation of choosing and approving

---

* Remember that any major investment in an IT platform or system may have serious ripple effects throughout the current IT complex. For example, it might require an upgrade to the organization's server or desktop platforms or the retraining of IT personnel. It also might require the renegotiation of existing software licensing agreements, a significant increase in maintenance and support overhead, and so forth. For these reasons, the business side of the house must be briefed on the TCO of any major IT venture before it commits. Also bear in mind that your TCO calculations must go well beyond first-time costs. They must include ongoing support and maintenance costs for the solution, as well.

what the IT unit undertakes, it is more likely to invest the time of its own people in such efforts.

Not all project suggestions will win the same level of support from business managers. Some will advocate for those that appear to benefit their own lines of business and may actively oppose those of neutral advantage to them. They will most certainly resist items that imply a need for their unit's support without a clear payback. Such responses could be viewed as self-serving, narrow, and petty, but they are also part of human nature. Not everyone shares the same agenda. In addition, your business colleagues may not fully appreciate the enterprise's overall operational needs or its real fiscal and technological constraints. It is therefore essential to find a way to build consensus around a prioritized list of IT investments for the coming year. This requires an approach whereby business leaders agree on the metrics for a balanced evaluation and ranking of proposal options.

Unfortunately, due to the complexities inherent in most IT projects, few in management may be able to discern either the true opportunities or the actual risks associated with a particular item on the wish list. The IT team must ensure that decision makers possess the necessary facts to make an informed choice. There are several dimensions to such deliberations. On the one hand, decision makers must understand each opportunity's value to the enterprise. On the other hand, they must appreciate what it will take for the business to realize that opportunity. In other words, they must know the effort's true scope in terms of cost, difficulty, and the potential for failure. To objectify this assessment of options, some practitioners develop simple scoring mechanisms based on a set of values and metrics agreed to by the decision makers themselves. Typically, such tools plot the ability of the organization to deliver the envisioned IT solution against the value of that solution to the enterprise and its customers. The end product of this process is a portfolio of projects that represent efforts of the highest value and with the greatest likelihood for successful delivery.

My colleague Tom Murphy has employed such a tool to great effect in both for-profit and not-for-profit corporate settings.[*] He observes that there are many benefits to his portfolio approach, including the following:

- Evaluation of all potential IT investments on the same set of criteria
- Customization of scoring factors to meet the unique characteristics of a particular company or institution

---

[*] Tom Murphy, "A Portfolio Planning Approach for IT Investment," *Enterprise Operations Management Journal*, 42-10-50 (August/September 2003): 1. For more information about the Murphy tool and related services, contact Tom Murphy at tommurphy25@aol.com.

- Balanced conversation concerning both infrastructure and business IT investments
- Creation of a commonly understood value proposition for IT undertakings that everyone across the business supports
- Identification of ways to improve the success of those IT initiatives chosen through the process

As Murphy also points out, the real teeth in his tool, or any tool like it, are in its ability to produce a standardized score for each proposal, based on business management–defined value and risk metrics. Decision makers may then compare these IT investment option scores and select those that possess the greatest promise of enterprise value while offering the least amount of risk in implementation.

The measures for a particular enterprise will no doubt reflect the values, concerns, strengths, and weaknesses of its business. For example, time-to-market or capturing a larger market share might be particularly important business objectives, whereas strong database and project management skills within the IT team might mitigate certain project risks. Through exploration and negotiation, IT leadership must construct the specific measures that apply to your situation. A generalized application of Murphy's process serves to illustrate.[*] See Exhibit 2.

In this example, the X axis of the table addresses project risk while the Y axis addresses customer value. Each measure has an upper point value with the total number of points assigned to both the X and Y variables equal to 100 points. The proposal risk factors include program scope and complexity, the strength of sponsorship and line-of-business commitment, the ease of end-user adoption, the quality and availability of appropriate skills and experience within the IT team, the IT team's project management capabilities, and the overall approach to project risk reduction.

Each factor is further defined by a series of accompanying definitional factors that are shown juxtaposed with each success or risk criterion. Thus, program scope and complexity may be understood in terms of the duration of the project, the breadth of users impacted by its implementation, the geographies or distances involved in the project's rollout, the level of IT integration required, and the complexity of the system functionality accompanying delivery. In this manner, the tool exposes all of the different aspects of each proposal on the enterprise wish list. It provides a uniform way to assess the organization's ability to deliver on that particular commitment. If not enough is known about the proposal to measure its inherent risks, this should speak volumes to those sitting in judgment on corporate IT priorities.

---

[*] Ibid., 3.

Exhibit 2 An Example of Tom Murphy's Portfolio Scoring Tool for IT Projects

Success Criteria Weights	Major Success Criteria	Specific Questions	Value Criteria Weights	Major Value Criteria	Specific Questions
X1 26 pts	Program scope and complexity	Duration Breadth of users Geographies IT integration Functions	Y1 30 pts	Customer advantage factors	Top 10 focus Category strength Industry response Ease of business Collaboration
X2 22 pts	Sponsorship and commitment	Support level Signoff Resources Funding	Y2 22 pts	Consumer marketing gains	Promotional gain Market insights Marketing mix POS data capture
X3 18 pts	End-user adoption	Involvement Resistance Ease of use Readiness	Y3 18 pts	Financial benefits	Inventory management IT cost savings Manufacturing efficiency IT revenue gain

Alignment and Planning — Doing the Right Things ■ 71

X4 16 pts	IT resource skills and availability	Dedicated team Consultant use IT skills Past experience	Y4 12 pts	Internal efficiency and production	Workflow Collaboration Functional Systems
X5 10 pts	Project management ability	Project plan Milestones Cost controls Project team Autonomy	Y5 10 pts	Strategic goal enhancement	Decision making Competitive edge of technology Global integration
X6 8 pts	IT risk reduction	Regulatory compliance Risk analysis	Y6 8 pts	Technology fit	IT integration Systems Network Applications
Maximum Points	100		Maximum Points	100	

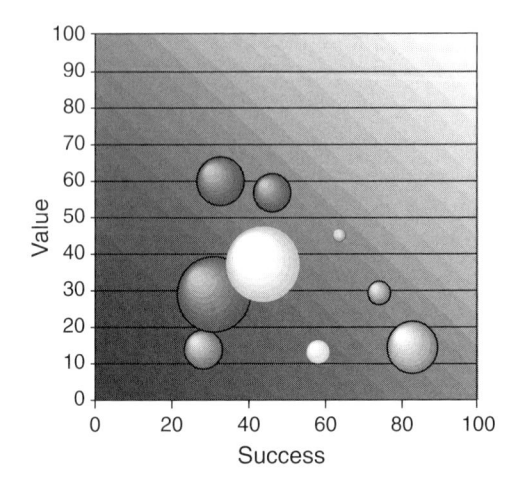

Exhibit 3 Murphy Tool Output — Potential IT Projects' Size versus Value versus Risk

This Murphy tool illustration also systematizes the estimation of project customer value in terms of tangible customer benefits, such as the attainment of known customer requirements; the growing of the business' market share; financial gain to the enterprise; internal improvements in business efficiency and work productivity; progress toward the realization of corporate strategic goals; and the technology fit of the solution. Here, too, the tool provides more descriptive details to assist the management team in scoring each factor. By applying the scoring tool as a group exercise facilitated by PMO project managers, the business can readily establish an objective measure of the risk and value for each proposal on the enterprise's IT project wish list. The results of this analysis may then be represented graphically for the purposes of comparison. See Exhibit 3.

For this process to work, both business and IT must get involved. Most of the information required to facilitate the score process may not be readily known. Joint business and IT teams must spend time in research and analysis, and further time will be consumed in meetings with management to reach consensus on what the data means in scoring particular proposals. Some subjectivity will undoubtedly creep into these discussions. By and large, however, the process should yield a reasonable level of informed, standardized assessments. As part of these deliberations, IT management must take every opportunity to clarify the real (as opposed to the hoped-for) costs of project delivery. Make it crystal clear up front that the true TCO for an initiative will require related IT costs; do not wait for these to surface during implementation. If the TCO for a particular proposal is unacceptable to management, resize the project to bring its costs back in line with the corporation's means or drop it from further consideration. These TCO calculations will not come easily and must be defensible. Here again, it is a good idea to let the PMO team do the research and provide an objective, independent opinion of analytical outcomes.

With this data in hand, leaders will be positioned to come to closure on a short list of approved IT projects. They may opt for those of highest customer value that also carry the greatest possibilities for successful implementation (upper mid-left hand quadrant of Exhibit 3.8), or they may choose fewer but riskier investments that offer great returns (upper right hand quadrant of Exhibit 3.8). Whatever their strategy, ensure that the ultimate decisions are left to the business leadership. For its part, the IT team must clarify the implications of one choice over another, especially when tradeoffs are required in the existing base of IT products and services. To this end, Murphy's approach, or something like it, will assist in leveling the field and forcing rational discourse. If the process is applied effectively, management will first agree to the process and the measures for prioritization and then will work with IT leadership to employ this model in making choices. Thus, this process ensures that the IT organization's resource commitments are aligned with the business objectives of the greater enterprise, taking the onus of saying "no" off the shoulders of IT personnel, allowing the IT team to focus instead on project execution and delivery.

As a result of this alignment process, the IT organization receives its directions for the coming fiscal year. IT management must now incorporate the surviving initiatives from the alignment process into its portfolio of ongoing activities. Ancillary maintenance and support efforts must also be incorporated into the plan, as must any tradeoffs in current offerings. The concluding section of this chapter explores how to capture these decisions as part of an annual IT planning and reporting process. Like the Murphy tool, the author's IT action plan affords a model that the reader can adapt to meet his or her particular needs. In all my years as an IT consultant and manager, I have found that the challenges — such as alignment with the business, balancing the demands for resources, and reaching consensus on the scope of work — are quite similar from one IT organization to the next. My processes and tool recommendations help the IT team to focus on what is most important in communicating more effectively its plans and results to customers and partner providers.

DOCUMENTING AND ACCOUNTING FOR IT PRIORITIES: THE ACTION PLANNING PROCESS

The information that comes out of the alignment and prioritization process is fodder for the creation of an enterprise IT planning document. Like many of the tools presented in this volume, the IT action plan serves any number of purposes.* First and foremost, through creating an IT action plan, the IT team can systematically survey, categorize, and reflect annually on all commitments in its portfolio of projects and services. As part of this effort, your IT team can adjust the scope of particular projects and service level delivery standards to ensure a proper balance between available IT resources and team commitments. Second, the IT action plan affords the formal, documented alignment of IT work with enterprise goals and objectives. Third, it serves as a vehicle for assigning responsibility within the team for specific IT deliverables, including scope definitions, specific delivery timeframes, and the metrics for customer satisfaction. Finally, it will most certainly assist the IT leadership and the team's customer relationship executives (if CREs are in place) in communicating across the IT organization and with customers concerning those projects and services of interest to them.

The underlying construct of the IT action plan is simple, albeit labor intensive. Its objective is to bring together information concerning long-term or strategic goals, near-term tactical objectives, and more immediate action items of the IT organization. Furthermore, this content must be structured and conveyed in a manner that makes clear the relationships between goals and objectives and between objectives and action items. Think of these elements relating to one another like the levels of a pyramid: the higher you go, the more generalized and abstract the plan's view of IT activities. Yet, all levels are connected: the more detailed (action item) layers align with and support the more general (strategic objective) top layers. See Exhibit 4.

As Exhibit 4 illustrates, the IT organization's mission statement rides over the entire portfolio of commitments and must embrace all IT goals and objectives. Conversely, each strategy must align in some way with the IT mission statement, and individual objectives and action items must line up under each goal. As a result, a good IT action plan will comprehend all that the IT organization delivers.

The process of creating an IT action plan should be highly participatory. This ensures broad support for the actual implementation of the plan. Also, given the complexities and interdependencies of IT service and project delivery, by involving a cross section of the team in formulating the plan, IT leadership will more likely avoid unfortunate omissions and false assumptions. The three sources of information that will help shape

---

* For an example of an IT action plan in use, see *The Hands-On Project Office*, http://www.crcpress.com/e_products/downloads/download.asp?cat_no=AU1991, chpt3~5~Action Plan~example. For the actual planning template, see chpt3~6~Action Plan~template. For a hardcopy version of this tool see Appendix B.

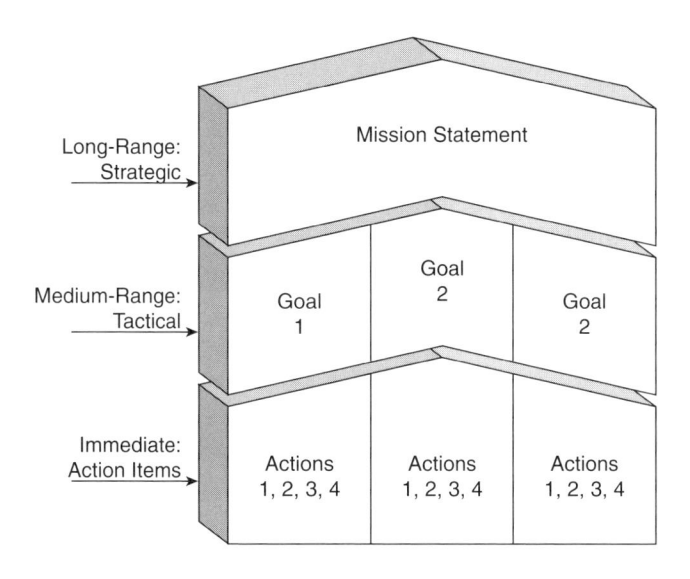

Exhibit 4   The Planning Pyramid

the formal planning discussion include the business-unit IT project prioritization process, the internal economy model and budget for IT, and your team's organization chart or some other source of clearly defined IT roles and responsibilities. All three of these documents will factor into the plan's ultimate formulation.

Depending on how your organization prefers to share responsibility, there are several approaches to creating the plan. Larger organizations may have a planning officer, such as the PMO director, whose responsibilities include collecting input and compiling a draft planning document. Alternatively, a small working group, facilitated by PMO personnel, might be assigned the task, or the CIO could take it on himself or herself. Once a plan is drafted, the preparers should walk through the draft with a representative cross section of IT team managers, subject matter experts, and the unit's CREs. Following this initial presentation and discussion, ask each reviewer to take a set period, say one week, to come back with questions, suggestions, and issues. Their input will help shape the document into its penultimate form, preparing it for a more formal review by IT management, including all the responsible parties whose names appear next to plan deliverables. Remember that the final product must align with the business, define deliverables and the parties responsible, and clarify how results will be measured. This is a lot for the team to digest and accept. Take the time to build a solid consensus around the plan, and involve your PMO in working with the various IT teams contributing to the planning document.

In practice, the author has used the IT action plan to document overall IT team commitments and to communicate these inside and outside of IT. In addition, I have used the plan to reinforce with my IT team our focus on delivering value to the customer and other operating principles. Last but not least, I have leveraged the process to draw out from staff a list of critical success factors and to build agreement around how to measure success in serving the customer. All of this rolls together in the author's IT action plan document, that includes the following sections:[*]

- IT organization vision statement
- IT unit mission statement
- List of goals and objectives, with language that links these to the enterprise's goals and objectives
- Operating unit guiding principles
- Competitive profiling information
- Guiding principles
- Critical success factors
- Observations concerning funding
- Observations concerning metrics and satisfaction measures
- Goal-by-goal breakdown of key initiatives defined in terms of:
  - Responsible parties
  - Timeframes and resources
  - Performance metrics
  - Self-review (a listing of major deliverables)
  - Manager's review (a listing of issues, barriers to success, and the realization of major milestones)
- Glossary of key terms

Of course, not all planning documents must include all of these items. Some corporate cultures will value detailed information sharing, while others will just want to-do lists organized by business unit customer. Find the right balance in crafting your presentation. It is better to say too much than to leave either your own team or your customers making their own assumptions about IT commitments. Be explicit up front to avoid disappointment later. Let us consider each of the plan template components in turn. The reader may then decide what works best for his or her own organization and business context.

Vision and mission statements serve both external and internal audiences. Externally, these declarations communicate IT's place and

---

[*] See *The Hands-On Project Office*, http://www.crcpress.com/e_products/downloads/download.asp?cat_no=AU1991, chpt3~6~Action Plan~template.

value proposition to the greater enterprise. Within the IT organization, they provide a sense of common identification and purpose. Of the two, vision speaks to and aligns with the enterprise's own mission statement. For example, if the corporate vision calls for world-class customer support, then the IT vision might say, "to assist the enterprise to achieve world-class customer support through the innovative use of information technologies." The IT unit's mission statement must be brief, while focusing more inwardly on what IT itself should be. Here is an example:

> To enable learning, teaching, research, and administrative services for the extended XYZ University community through the effective and efficient deployment and support of proven information technologies and the best IT business practices commensurate with the highest levels of customer satisfaction.[*]

Like all good mission statements, the example addresses the following in general terms:

- Primary business focus of the enterprise
- IT's customers in that context
- What IT commits to deliver
- How the results are to be measured

As such, it is a powerful, galvanizing statement for staff and customers alike. Although the vision and mission statements are brief, they are not to be taken lightly. The IT organization will hang its collective hat on these statements. They should resonate within the team and with IT's customers and partner providers. Spend time developing these statements and ensure that your team is fully behind them before you take IT's vision and mission public.

By their nature, mission statements — no matter how exciting — are broad expressions of intent. To give these global declarations depth, create complementary IT unit goals. If properly structured, four to six goal statements ought to comprehend all IT deliverables. Obviously, the text will vary with the particulars of your business. Here are a few general rules:

---

[*] This example is drawn from the sample action plan, http://www.crc-press.com/e_products/downloads/download.asp?cat_no=AU1991, chpt3~5~Action Plan~example.

- Construct no more than one goal per line of business served; if you do the same things for several lines of business, consolidate them under one goal.
- Employ a single goal for all infrastructure-related projects, i.e., initiatives sponsored by IT itself that provide and maintain the enabling technologies for more customer-centric IT products and services.
- Devote a goal to IT organization process improvement and the pursuit of best practices, i.e., the running of IT as a business, including such activities as resource planning and budgeting, asset management, information security, customer relationship management, and the like.
- Create a goal that addresses the needs of the culture and people within the IT organization itself, e.g., staff recruitment and retention, staff training and development, performance management, team building, and so forth.

Each goal should identify in general language what is to be done, who is to be served, and how results are to be measured. Make a point of aligning each of your team's goals with the mission and goals of the enterprise. Simple but comprehensive goals are best. Again, try to keep the total number of goals within reason; employing too many goals will fragment and defuse the IT team's focus on deliverables. At the same time, ensure that your plan's goals, in total, encompass the sum of your team's commitments.

Drawing on these principles, a sample set of IT unit goals might include the following:[*]

- Goal 1 — to enable and support the University's learning, teaching, service, and research objectives, through the delivery of information technology products and services in keeping with available University resources. (This aligns with enterprise strategic goal x.)
- Goal 2 — to enable the administrative needs of XYZ students, faculty, and staff in an effective and efficient manner, through the delivery of information technology products and services, reflecting best higher education business practices. (This aligns with enterprise strategic goal y.)
- Goal 3 — to establish and maintain a robust, scalable, flexible, reliable, and cost effective information technology infrastructure that supports the current needs and that anticipates the future needs of the XYZ University community. (This aligns with enterprise strategic goals z and y.)

---

[*] Ibid.

- Goal 4 — to establish and maintain IT division policies and practices that protect the University's investments in information technology, business systems, and enterprise data and that respect the rights and privileges of all community members. (This aligns with enterprise strategic goals *a* and *y*.)
- Goal 5 — to strengthen the performance of the IT division through the recruitment, retention, ongoing development, and tools enablement of the very best information technology professionals. (This aligns with enterprise strategic goal *b*.)

In this example drawn from higher education, the lines of business served are collapsed into two goals: one in support of academic delivery and one in support of administrative operations. Goal three encompasses all infrastructure work; goal four includes all process improvement activities, and the final goal speaks to IT staff issues and needs.

The next series of action plan elements is optional and may or may not fit into the culture of your organization. Although the author strongly recommends inclusion of these elements in your plan, you may need to adapt them for your use. The first of these is competitive benchmarking data. Often, IT management is asked to defend its choices and expenditures in light of what other IT organizations offer or spend. This is not an unreasonable request on the part of your business colleagues. No doubt they are being asked to benchmark their performance, as well. What matters here is the selection of appropriate comparisons. Industry analysis firms, like Gartner and META Group, regularly offer IT spending survey data. Your organization may or may not fit within one of their survey categories. Study the survey criteria with care before offering a comparative study to business management as an approach to benchmarking your operation.

Although they may be hard to come by, appropriate company-to-company comparisons will also prove useful. Just make sure that the same metrics are used by your team and those with which you compare. Typically, one finds such data offered through professional associations, trade consortia, and the like. If you do pursue benchmarking as part of your planning and management processes, be sure to vet this list with your business colleagues and get their approval up front. Set expectations about what you and they can reasonably derive from these data sets, and then maintain the agreed-on process for data collection, comparison, and analysis.

Next, consider the inclusion of IT team operating principles. These are behavior norms governing the ways you and your team interact with each other and with customers. Team operating principles may prove to be a critical element in building a high-performance, service-oriented

IT organization, but only if they permeate your internal culture and your dealings with customers and partner providers. If you fail to acknowledge and reward those who exemplify adherence to these principles — and if you neglect to practice them yourself — they will fade into obscurity. On the other hand, if you continually reinforce their relevance to the success of the team and the health of the overall work environment, they will serve as a potent management tool. Here is sample list of IT operating principles:

- Prioritize for customer value
- Manage through consensus and deliver as a team
- Make and honor commitments, but make no commitments for others
- Go directly to the person or the problem to seek out the root cause(s) of problems and disputes
- Value collaboration and information sharing
- Respect each other and value difference
- Manage by fact, not by assumption, rumor, or personal agenda
- Take responsibility for the team's success
- Celebrate the journey and learn from the experience

Complement the use of IT operating principles, which are inwardly focused, with use of customer-oriented critical success factors (CSFs). CSFs focus on those business practices that ensure positive outcomes in interactions with customers and improve project and service delivery more generally. Like operating principles, CSFs address the soft, interpersonal side of transacting with your line-of-business colleagues. They also afford a framework for the operational maturation of the IT organization. CSFs should be concrete, focused on customer delivery, and leading to action, as these examples show:

- Communication
  - Clear, continuous communication, throughout the IT unit and with customers and external partners, that focuses on the clarification of roles, responsibilities, and deliverables
- Delivery management
  - Predictable delivery of products and services
  - Project and subcontractor requirements and IT architecture management that lead to timely and cost-efficient implementation and operation of the IT products and services
- Resource management
  - Resource flexibility, deployed and allocated to highest-value business requirements
  - Attraction and retention of top-performing staff who are aligned with business priorities

- Architected and managed solutions
  - Products and services, whether developed in-house or provided from external sources, that are well integrated with customer needs and XYZ's existing base of information technologies
  - Adherence to XYZ's IT architecture and engineering standards and the consistent use of quality assurance (QA) processes
  - Baselining and ongoing measurement of performance and regular reporting to the appropriate stakeholders

In employing CSFs, you are also reminding your IT colleagues that your business must remain focused on the needs of the customer and hence on service and project delivery.

If there are special financial considerations relevant to the IT Action Plan, you might include these as part of your planning document. For example, if your plan makes certain assumptions about capital funding, asset depreciation, or ongoing maintenance and support, clarify these in the document. Similarly, you might consider a brief explanation of the role that metrics will play in the management of the IT plan's implementation.

As the sample IT Action Plan shows, all of these plan components are covered in two pages. These pages provide a framework for performance. The remainder of the plan devotes itself to actual IT commitments. Here, the template is straightforward:

- IT initiatives are organized under the appropriate goal statements reproduced from page 1 of the template.
- Each initiative is an umbrella statement under which the team might position any number of more discrete IT activities.
- In total, the initiative statements should capture all IT team activity — service and project delivery.
- To the right of each initiative, indicate the primary and secondary parties responsible for actual delivery, mostly team leaders and subject matter experts, but also support personnel from the PMO.
- Next, identify key milestones and dates.
- Provide performance metrics for each deliverable. Note that these are results and not activity measures. You need to ask yourself, from the customer's point of view, what reflects satisfaction with the IT deliverable in question, measuring and reporting plan results based on these metrics.
- The remaining two columns are completed each time the plan is updated:
  - The self-review column affords a place for the capturing of specific accomplishments over the last quarter or year.

- The manager's review column allows for executive-level comments concerning the nature of the accomplishment, barriers confronted and overcome, and next steps or changes in plan.

Taken together, this information provides a comprehensive yet concentrated view of IT commitments and accomplishments over a given period. The document clearly identifies who is responsible for each IT deliverable and how performance will be measured. Overall, it sends a powerful message to the team about customer-focused work priorities.

IT management will want to share this document widely within the organization and most probably with corporate sponsors (i.e., funding sources). Because some of the language employed in the plan may be obscure to those not involved in its creation, add a brief glossary of key terms at the end of the plan. Also, consider customized views of the plan for targeted enterprise audiences. Remember that through this document, IT strives to communicate more effectively the nature and extent of its commitments to the constituencies served, and to ensure the understanding of and agreement to these commitments among IT operating units and personnel. Typically, a plan will be reviewed and updated quarterly to capture service delivery and project accomplishments, as well as issues impacting future deliverables. Adjustments to the IT action plan may occur throughout the year. These changes may be driven by the dynamics of the marketplace, evolving business requirements, or emerging IT opportunities.[*]

The final section of this chapter focuses on the overall annual process for plan creation, leaving the consideration of off-cycle plan adjustments for Chapter 4 and Chapter 5. Your own organization's annual planning and budgeting processes may be rather complex, with numerous cycle events. For example, the business side of the house may have established procedures for strategic planning, capital allocation, staffing, project approval, and so forth. To achieve properly balanced decisions between project and service priorities with funding allocations, the IT planning process must integrate with these established activities.

To illustrate the dynamics of the situation, I have provided two process views. The first of these is a planning timetable spreadsheet,[**] which illustrates the timing and interrelationships between business and fiscal year cycles,

---

[*] If a proposed change to the plan will require a significant change in service delivery levels, impact multiple IT operating units, or involve additional IT resources above a certain threshold amount (say, $25,000), the process should accommodate such adjustments. Chapter 5 considers these issues in more detail.

[**] See *The Hands-On Project Office*, http://www.crcpress.com/e_products/downloads/download.asp?cat_no=AU1991, chpt3~7~planning~timetable~example and chpt3~8~planning~timetable~template. For a time line format and work flow for your planning process see chpt3~9~annual planning process~work flow.

strategic planning, performance management,* budgeting planning, capital planning, and so on. One cannot hope to manage what one does not comprehend. By mapping all of your organization's planning activities next to each other, you can identify more readily those that require input from the IT management team. Where there are gaps in the process, you can introduce changes, and where timing is critical, you can focus staff resources to deliver what is needed when it is needed. The planning timetable template also may serve as a useful educational tool to prepare your IT colleagues for the coordinated effort of developing plans, schedules, budgets, and investment estimates in line with the requirements of your line-of-business partners. This timetable would serve as a to-do list for those PMO personnel supporting the IT organization planning and alignment processes.

It is also helpful to create a flow chart of the internal IT process, with its iterative steps and key delivery dates. As modeled in Exhibit 5, a representative process is fed by many sources of information beyond those of the lines of business, including needs and opportunities identified by the IT staff and forces active in the external marketplace. All of this data is then filtered by IT management to terms of its impact on the following:

- Existing service level agreements (SLAs)
- Established and ongoing IT projects
- Embedded IT infrastructure of the enterprise
- Quality and size of the organization's IT work force
- Established IT funding levels

My model is iterative at this point because analysis may call for additional information gathering. PMO personnel will then bring together its documentation for reviews within the IT management team. Eventually, the content will be ready to share with appropriate members of enterprise management. See Exhibit 5.

As the larger organizational processes grind on, IT must prepare for action. Here, the workflow suggests two separate but interrelated streams of activity. On the one hand, IT management will continue to interact

---

* Although performance management is not a particular focus of this book, incorporating the values and metrics of the IT action plan into IT performance management is essential to the health of both processes. The IT team will adhere more strongly to the operating principles and delivery commitments if team members understand that continued employment and pay increases are tied directly to the modes of behavior and deliverables represented in the IT plan. For a performance review template that does just that, see *The Hands-On Project Office,* http://www.crc-press.com/e_products/downloads/download.asp?cat_no=AU1991, chpt3~10~performance review~template. This and similar tools should be maintained by the PMO for IT organization management.

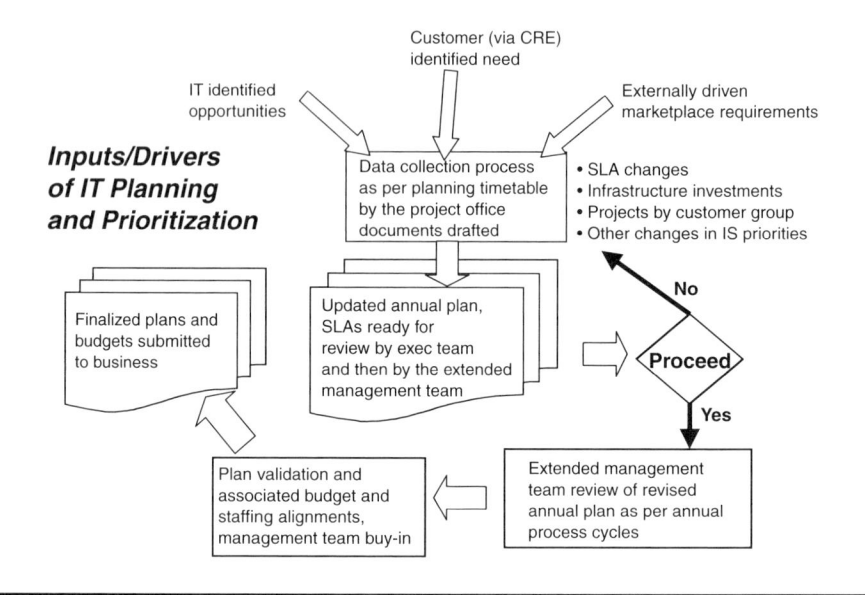

Inputs/Drivers of IT Planning and Prioritization

- IT identified opportunities
- Customer (via CRE) identified need
- Externally driven marketplace requirements

Data collection process as per planning timetable by the project office documents drafted

- SLA changes
- Infrastructure investments
- Projects by customer group
- Other changes in IS priorities

Updated annual plan, SLAs ready for review by exec team and then by the extended management team

Proceed — No / Yes

Extended management team review of revised annual plan as per annual process cycles

Plan validation and associated budget and staffing alignments, management team buy-in

Finalized plans and budgets submitted to business

Exhibit 5   The Action Planning Process Workflow — Part 1

with business colleagues on plan refinement. On the other hand, the IT organization, through its PMO, must initiate SLA reviews and project plan development in anticipation of the initiatives scoped in the documents under review. Any changes to what was originally proposed must be sized and assessed against resource allocations. In a well coordinated process, the IT organization will be poised for action when the plans for the coming fiscal year are finalized. See Exhibit 6.

Although this sounds plausible, the author's model is somewhat optimistic. Even the best corporate planning processes do not always produce clear direction. For that matter, some fail to follow their own timetables, leaving the resolution among choices for some time in the new year. When this occurs, IT is left in a holding pattern, only to find itself with squeezed delivery dates or project log jams. Still other organizations make their IT investment decisions on a revolving calendar basis, releasing resources along with assignments. The reader should adapt his or her sense of the process and the tools provided here to accommodate the realities of his or her workplace.

In the end, the success of any IT alignment and planning effort has less to do with the rigor of the process and more to do with the quality of communication among participants. In these discussions, the contributions of the IT leadership involve the timely delivery of relevant information and the realistic shaping of business leaders' expectations. Reason may prevail, but fantasy may pervade the alignment and

Alignment and Planning — Doing the Right Things ■ 85

```
┌─────────────────────┐
│ Finalized plans and │
│ budgets submitted   │
│ to business         │
└─────────────────────┘
           ▲
           │
┌──────────────────────────┐    ┌──────────────────────────┐
│ Plan validation and      │◄───│ Extended management      │
│ associated budget        │    │ team review of revised   │
│ and staffing alignments, │    │ annual plan as per annual│
│ management team buy-in   │    │ process cycles           │
└──────────────────────────┘    └──────────────────────────┘
```

- In May exec team to adjust plans based upon final budget allocations; assign project directors
- In May/June planning results communication, plans adjusted, resources
- In June extended management to scope work and finalize prioritization
- In June PMO to assign project manager/B analyst; work proceeds as planned
- In July, performance review docs reflect annual plan
- Quarterly progress reviews: September, December, March, and June

```
┌───────────────┐    Yes  ╱Changes╲  No    ┌─────────┐
│ Revisit and   │◄───Resource   Resource──►│ Proceed │
│ adjust plans, │    implications implications└─────────┘
│ get commitment│
└───────────────┘
    ▲
Restart
annual cycle
```

- If SLA activities, adjust SLA(s)
- If project, get sign-off on commitment doc and project plan
- Ensure that resources are assigned and prioritized
- As SLA work/project begins, scorecard created, tracking begins

---

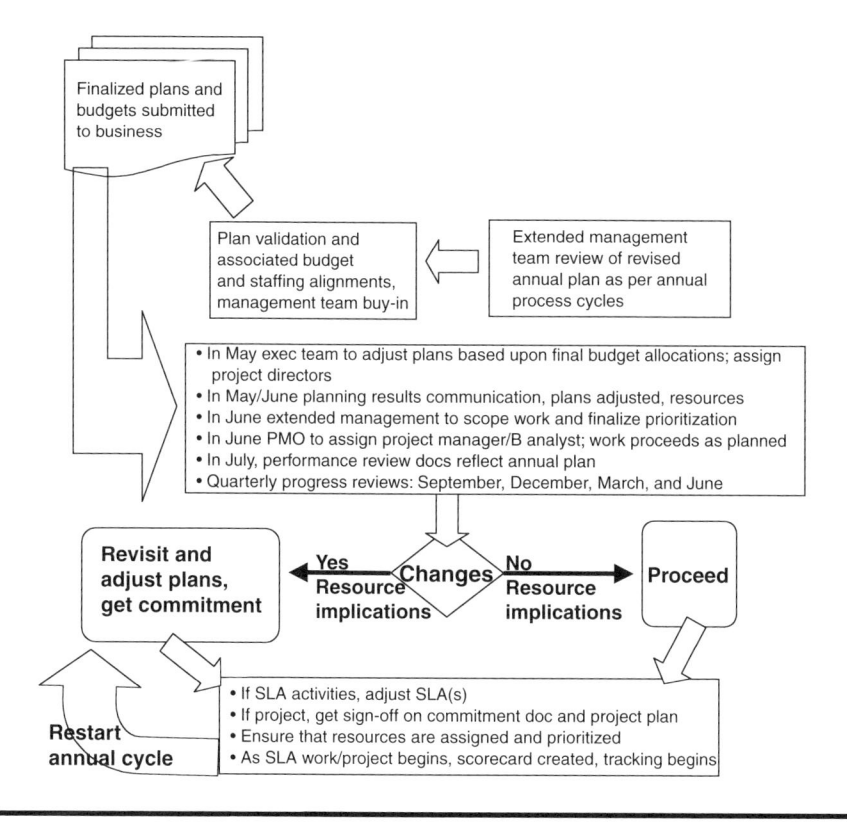

Exhibit 6   The Action Planning Process Workflow — Part 2

planning efforts. Stick to the facts. Do not promise too much, leaving your team in an untenable situation when it comes time to deliver. Rely, if you can, on the services of your PMO to coordinate process documentation, working session scheduling, and intra- and interunit communications. Include review sessions to ensure that the message reaching your customers is in line with your own thinking and that of your management team.

To ensure further the ongoing quality of these communications and to help your leadership team succeed more generally, I recommend the establishment of at least one IT advisory board made up of representative line-of-business partners from across the enterprise. For that matter, in a not-for-profit setting, I have employed two boards: one comprising key internal customers and the other drawn from the external community of institutional allies (e.g., alumni, retirees, former board members and officers), IT experts, and business partners (e.g., product and service providers). If well managed, these boards will provide honest feedback on your IT organization's plans and performance. They will also offer practical advice, fresh ideas and perspectives, and at times, even added resources to assist IT in better delivery and performance.* Furthermore, by actively employing an advisory group, the IT leadership team will have another way to communicate with those they serve. This investment of time and effort will pay off in many ways, including improved relations and understanding between IT and the rest of the enterprise, clearer alignment between corporate priorities and IT activities, and a focused sounding board for the launch of new initiatives.

Assuming a positive outcome from the planning process activities cited above, the IT organization must now bring to fruition those service and project commitments articulated or assumed in enterprise plans. But even the best-laid plans, especially when it comes to IT applications, can take unexpected turns. Although the aforementioned process prioritizes work and allocates resources over a timeframe of 12 months or even longer, it cannot anticipate changes in the business or in IT that might create more pressing needs or more promising opportunities. These so-called off-cycle events merit a management process of their own. This workflow may be illustrated as shown in Exhibit 7.

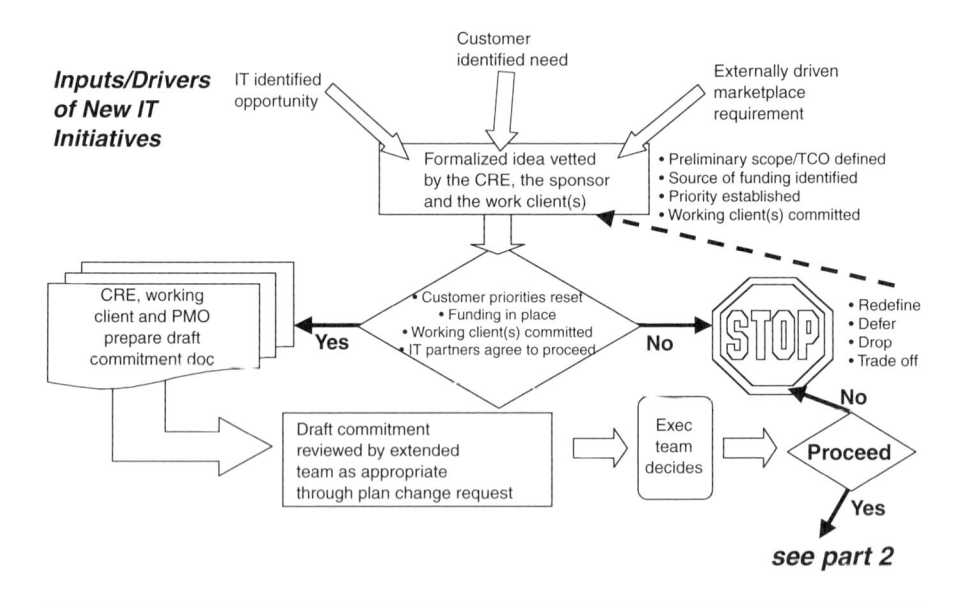

Exhibit 7 Off-Cycle IT Project Intake — Part 1

---

* For a sample IT advisory board charter, see *The Hands-On Project Office*, http://www.crcpress.com/e_products/downloads/download.asp?cat_no=AU1991, chpt3~11~IT Board charter~example.

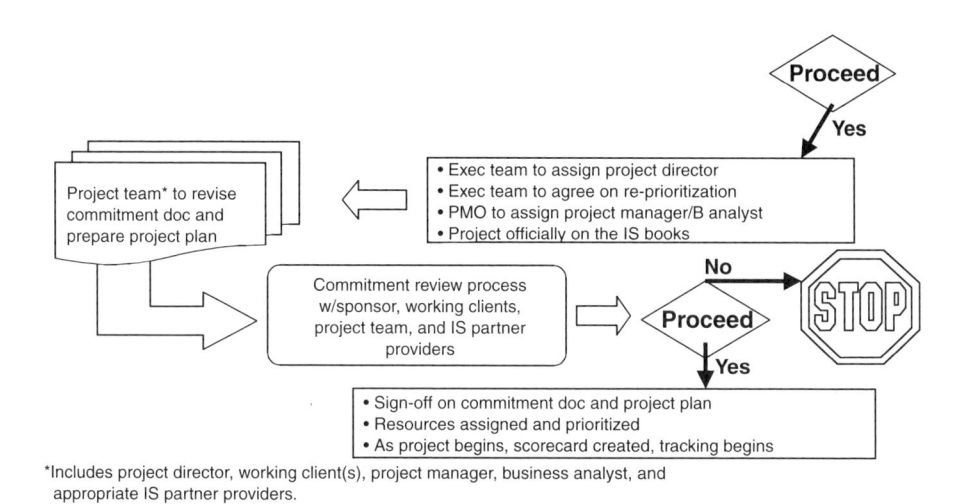

Alignment and Planning — Doing the Right Things ■ 87

```
                                                                    ◇ Proceed ◇
                                                                         │ Yes
  ┌──────────────┐       ┌─────────────────────────────────────┐         │
  │ Project team*│       │ • Exec team to assign project director│        │
  │ to revise    │ ←──── │ • Exec team to agree on re-prioritization│
  │ commitment   │       │ • PMO to assign project manager/B analyst│
  │ doc and      │       │ • Project officially on the IS books │
  │ prepare      │       └─────────────────────────────────────┘
  │ project plan │                                              No
  └──────┬───────┘       ┌─────────────────────────────────┐  ─────→ [STOP]
         │               │   Commitment review process     │         ↑
         └──────────────→│   w/sponsor, working clients,   │ ──→ ◇Proceed◇
                         │   project team, and IS partner  │         │ Yes
                         │          providers              │         ↓
                         └─────────────────────────────────┘
                         ┌──────────────────────────────────────────────┐
                         │ • Sign-off on commitment doc and project plan│
                         │ • Resources assigned and prioritized         │
                         │ • As project begins, scorecard created, tracking begins│
                         └──────────────────────────────────────────────┘
```
*Includes project director, working client(s), project manager, business analyst, and appropriate IS partner providers.

Exhibit 8   Off-Cycle IT Project Intake — Part 2

As in the annual planning process, each change to the plan must be thoroughly scrutinized. Does it have a sponsor and working clients? Is it funded? What are its TCO implications? What is the value of this proposal to the enterprise versus the values of those already approved and in the queue? If it does have a greater value to the enterprise, and assuming no additional resources or capacity exists within IT to get this additional work in the pipeline, what project(s) will be dropped from the original list, and is there consensus around that choice?* If a proposed addition to the annual plan passes these reviews within the IT organization itself, it may then proceed to the next level of the intake process, where it is vetted by the PMO through a more detailed commitment process and project plan. See Exhibit 8.

During this phase of the process, the PMO will assign a project manager, and perhaps a business analyst, to create the necessary scoping documents. Once these are ready, the package must go before stakeholders for final approval before it enters the IT organization pipeline. If the tradeoffs involve two projects owned by the same sponsor, this could be a fairly easy conversation. But if, as is often the case, one sponsor must forgo a project so that another business unit may pursue one of its own, the conversation could be heated. Be sure to present IT's position in a neutral manner. Stick to the facts and rely on the

---

* For an electronic version of this workflow, see *The Hands-On Project Office*, http://www.crcpress.com/e_products/downloads/download.asp?cat_no=AU1991, chpt3~12~Off-Cycle Approval Process~workflow.

process' value measurement tool to identify the superior opportunity. Let business management make the final decision. When you are faced with a pure add (i.e., no tradeoffs, just added work), you may need to revisit the scope of other approved projects to accommodate the demands of the new project. Here again, use the established rules of the process and its tools to present your case, and allow rational discourse and sound business thinking to prevail.

In today's rapidly changing world, the adaptability of IT's management processes is paramount. Do not assume that the plan with which you conclude the formal planning cycle is cast in stone. Flexibility and responsiveness to evolving business needs are the hallmark of a winning IT organization. The remaining chapters of *The Hands-On Project Office* explore in more detail the organizing principles, work processes, and best practices that can help the reader's IT team achieve these positive results. Chapter 4 discusses the particular approaches and tools for service delivery management. Chapter 5 does the same for project management. These two chapters speak to the proper conduct of work stemming from and driven by the IT action plan.

4

MODELING AND MANAGING SERVICE DELIVERY

INTRODUCTION

Ask any chief information officer (CIO) what he or she is doing to contribute to the success of the enterprise, and the CIO will respond with a list of all the strategic IT projects currently under way. However, if you were to poll that IT executive's customers, a different line of discussion would emerge. Although some customers might well focus on the recent move to E-commerce or the introduction of an enterprise resource planning system, most would talk with great intensity about network availability, system response time, the quality of help desk personnel, and the overall reliability of the organization's hardware and software platforms. In short, the typical corporate customer values IT as a utility: always there, always on, and as dependable as your wristwatch. Indeed, those IT organizations that are highly regarded by their parent institutions have earned this standing because of their attention to and success in day-to-day service delivery.

Excellence in service delivery comes to those IT teams that commit energy, resources, and focus to the seemingly mundane yet essential management components of IT performance. Other chapters in this book consider IT planning and alignment processes, organizational design, and project management best practices. The purpose of this chapter is to provide a series of frameworks for the effective positioning and communication of service delivery management within an IT organization and between that organization and its customers. Although excellence in service delivery is the responsibility of the entire IT organization, good housekeeping practices require their own champions. IT management must serve as these champions, while the PMO provides operational assistance and support. With that in mind, the chapter offers a series of illustrative models for day-to-day oversight and measurement of IT performance in the mission-critical areas of customer care and satisfaction. Here, too, the PMO can help the rest of IT maintain these practices.

To that end, it makes sense to say a few words about the concepts underlying IT service delivery management and the extent to which the reader's organization embraces these values. Let me begin with the work principles for the design of service delivery processes.* These are guiding principles that I employ in modeling service delivery scenarios for my own IT organization. First, I ask myself, "What should my customer experience in interacting with IT?" Then I consider how I would like my IT colleagues to view their roles in these service delivery scenarios. Your PMO team is properly positioned to discover what principles and associated scenarios will work for your IT organization. To facilitate this self-examination, adapt this series of value statements to set targets for your own IT team's performance:

- Single window/seamless service — services will be delivered in a customized manner to all customers through an integrated, single window, a so-called single point of service. Such an environment will provide the customer with a sense of personalized, focused, and responsive attention while achieving economies of scale and efficiencies in performance for the enterprise as a whole.
- Streamlining — all processes between the customer and the enterprise's IT service providers will be minimized through the elimination of unnecessary steps, forms, and procedures; the reduction of management layers; and the provision of more direct contact between service providers and customers.
- Choices — customers will be given choices about services and how these are delivered. Flexibility, responsiveness, and choice will become the hallmarks of reengineered IT business processes.
- Consistency — within the IT unit, and between the IT team and its enterprise customers, the same processes and services will be conducted and delivered in the same way, eliminating unnecessary complexity and its associated costs.

---

* Although many of the values expressed in this chapter have surfaced during the author's years as a change manager, the organization and language employed here come in part from Secretariat, Treasury Board of Canada, *Blueprint for Renewing Government Services Using Information Technology* (Ottawa, Ontario: Treasury Board, 1994), discussion draft. My thanks to Sue Gavrel and her colleagues for their assistance in making this draft document, which is important as a blueprint for current-day thinking about electronic service delivery, available.

- Location and time independence — thanks to IT (and where practical and cost justifiable), customers and service providers will have access to information and services at any time and from many places.
- Continuous improvement of service — processes and services will be improved on an ongoing basis. The customer-driven measurement mechanisms embedded in these service processes will provide the basis for their continuous improvement. These same measures will afford the quantitative, as well as qualitative, means of assessing IT's value to the greater enterprise.

Given these customer service objectives, the reader can perhaps envision how enterprisewide IT operations might be transformed. Indeed, my illustrations may suggest service delivery scenarios that are different from the way your team currently thinks of its role within the enterprise. Although your team's focus must be on the customer and how best to serve his or her needs, your customer is also a partner in, not the lord and master of, IT service delivery. The values of convenience and ease of use must be weighed against practical cost and operational considerations. The key to success is to find the right balance. Solicit feedback from all sides. In the end, however, your management team must make informed choices and live within the constraints imposed by your business, your budget, and the technologies at hand. The following list of customer service scenarios is not definitive, but it does illustrate some possible outcomes of applying these values and principles to actual IT-enabled business process scenarios:

- Automatic service — in this service scenario, a customer's own computer system generates a service request, and the supplier's system provides a response, with minimal human intervention. For example, an inventory control system might identify the depletion of a product from warehouse shelves and automatically reorder that product from the appropriate supplier. Although human intervention is required in the review of system rules, inventory-level parameters, and the like, once these rules are in place, the system will run on its own and at little ongoing cost to the enterprise.
- Self-service (electronic) — this mode of service delivery has come into its own in recent years. Customers use workstations, kiosks, or telephones to access information and to generate transactions, orders, and payments, resulting in reduced or eliminated paperwork and fewer approval steps. Automated teller machines, banking and investing by phone, and more recently shopping via the Internet are well known examples of IT-enabled self-service at its best.

- Self-service (walk-in) — Wal-Mart®, Home Depot, Inc.®, and catalog distribution centers have demonstrated the success of this service delivery model. In it, customers seek information, goods, and services by visiting common walk-in centers, where service providers use computerized systems to respond effectively and efficiently. Of course, government agencies may be the counter-proof of this model, but for some applications it clearly works. The single-point-of-service strategy is particularly attractive in settings where the processes are complex or the choices are many. For example, the delivery of academic and administrative support services in colleges and universities lends itself to this type of design.
- Service with on-site support — at times, the most effective mode of service delivery is on-site. In this circumstance, local experts provide multiple services for multiple customers and cross-functional service providers, maximizing the benefits of information technology and minimizing duplication and paperwork. Many aspects of employee training and development, information resource management, and IT support lend themselves to on-site servicing, where a knowledgeable person may respond in a customized and timely manner to situations as they arise.
- Specialist/expert service center — at times, the problem or the customer requirement in a particular service setting demands the attention of a specialist. It would be economically and logistically impossible to provide such a person physically at each location and for each instance when such expertise is required. Fortunately, through the enterprise's information network, service providers and their customers can access experts — both human and computer-based — directly and quickly, reducing the need to duplicate similar services and improving the rate and success of customer response within an economical framework. Consulting firms, engineering firms, law offices, hospitals, and high-tech companies all typically leverage expertise in this manner.
- Supplier interfaces (entire enterprise) — finally, one can envision a service delivery scenario in which suppliers and customers are connected directly to each other. Such an arrangement, in the case of missing parts, out-of-stock items, or subcontracted expertise, can lead to prompt responses to customer needs without the added time and expense of obtaining assistance from the distributor or general contractor of the service. For this model to work, suppliers and customers must be viewed and view themselves as partners in a business process value chain. Today, such practices are commonplace through Web-enabled, business-to-business, E-commerce transactions.

Which of these six settings applies to your organization, and what are the implications for IT service delivery? What kind of service experience does your customer need or expect? Perhaps all or just a few of these examples will resonate. In all likelihood, they serve best as representational models for you to mix and match, as you and your team think through your customers' requirements. Because service delivery is such an important measure of IT organization contribution and value, these conceptual frameworks are useful as guides in scripting the more in-the-trenches view of process management that must follow these musings. Even as we explore the more mechanical processes of setting, employing, and reporting on service delivery performance measures, keep in mind the big picture of how your IT team is perceived in its service delivery roles.

MODELING SERVICE DELIVERY MANAGEMENT

Every type of enterprise today requires some level of IT enablement. The vendor community, and at times IT leadership, tend to oversell the value and understate the required investment and difficulties in bringing enabling technologies to bear. In truth, enterprisewide IT deployments are complicated and resource intensive — at the outset and in terms of ongoing maintenance and support. Indeed, once IT solutions are operational within an enterprise, they must run continuously, with the IT organization adjusting on the fly to added users, new user requirements, and changes in the direction or the nature of the business. Unfortunately, too much tends to be promised by the IT organization and its external partner providers, and too much is assumed by the end-user community. These circumstances can lead to misunderstandings, disappointed expectations, waste, and even poor performance results. On the flip side, too often good services come through heroic effort rather than a routine, predictable, and manageable process.

To better align end-user expectations with IT organization capabilities, many IT executives employ customer relationship executives (CREs), as described in Chapter 2, to serve as the voices and the ears of the IT team in dealing with its business unit counterparts. To complement this personalized approach to communication, some IT organizations also employ service level agreements (SLAs), as briefly described in Chapter 1, to communicate with their customers and to educate them on the nature and extent of the relationship between the specific business units and IT services. Together, the CRE and SLA processes help shape and inform customer expectations about service, thus functioning as a safety valve in a well-managed delivery process. To achieve their operational objectives, both are supported directly by PMO personnel.

The interactions between the CRE and SLA processes are easily described. First, IT management must identify its key customers, typically the senior executives of the business or operating units served by the IT organization. Second, the CIO or his or her designate should visit with each executive and his or her management team. In these preliminary sessions, the senior IT executive outlines the intentions of the service delivery management process, reviews the actual SLA form (described below), and considers an appropriate CRE to serve as a liaison with that business team. As an outcome of this effort, each business unit's leadership will have some sense of the IT organization's service level management process; the scope, nature, and current capabilities of IT in the service of the business unit; and the role of the IT CRE who will serve its interests.

The CRE will then help to prepare a complete SLA for the review of his or her assigned business unit. To do so, the CRE will turn to the PMO for information concerning the history of that customer's prior year of work for that business unit. The CRE will also request a catalog of all the products and services delivered to that business unit during the course of the year, as well as their associated costs. In an organization in which the business unit is paying directly for IT services, accurate cost and service delivery information is essential. In organizations in which IT is funded directly by the enterprise to provide IT services, however, cost information may only confuse the issue. The SLA process and form should fit the particular circumstances and culture of your enterprise.

Last but not least, the CRE works with the PMO and key IT service managers to establish all the appropriate IT service level metrics that quantify IT performance as these relate to the business unit in question. This data will be employed both in the SLA document and in conversations with business unit leaders to establish mutually agreeable service delivery expectations.

Once the CRE, with the assistance of the PMO, has done his or her due diligence within the IT organization to prepare the SLA, the document should be reviewed carefully while it is still in draft by the CRE's business unit counterparts. After the draft has been properly vetted, the CRE will more formally present and review the SLA, which now serves as a service delivery contract, with business unit management. Again, given the culture and internal business practices, the nature and form of this contract will vary from enterprise to enterprise. Thereafter and on a regular basis (e.g., monthly), the CRE will sit down with his or her business unit customers to discuss how well the IT organization is performing against the metrics set forth in the SLA. During these discussions, the CRE will also listen for opportunities to improve or enhance existing services and for new business opportunities (e.g., product or service enhancements and projects).

Because the SLA includes actual performance metrics, these metrics will provide a quantifiable basis to assess the quality of IT service delivery. In large IT organizations, problem ticket systems are often employed to monitor the status of problem resolution (i.e., repairs to existing IT services) and service requests (i.e., customer requests for access to or enhancements of existing IT services).[*] These systems track the history of a customer ticket from its point of entry, usually through the IT help desk or call center, to the point when it is closed (i.e., when the customer's needs have been addressed). To work properly, any tracking system must be built on a foundation of agreed-to workflows and handoffs. The standard for each step's delivery is time-based and assumes satisfactory results at the close of each step in the process. Typically, these rules and metrics are captured in the system, and actual performance is then recorded and compared to these metrics. To facilitate this effort, I suggest modeling these processes in PowerPoint or Visio and then marketing them across IT to establish a consensus on their accuracy and their practical implementation.[**] In each instance, your help desk people would log the problem incident or the service request into the tracking system before passing it on to an IT service provider for processing. By way of illustration, here are two performance metric standards. The first addresses problem resolution, and the second addresses service requests. See Exhibit 1.

In the problem resolution metrics table, help desk personnel create and rate problem tickets based on their perceived level of severity (i.e., the extent to which the customer and the enterprise are impacted). As this table suggests, the IT organization commits to getting back to the customer within a prescribed period of time after receipt of the problem ticket from the help desk; IT also commits to resolving that ticket within a given timeframe. Each metric is based on customer input and enterprise needs, but each is also based on the history of past IT team performance.[***]

---

[*] Most organizations will set a threshold on the costs and risks associated with a service request. Any request that exceeds this hurdle rate (e.g., the project is either high risk or exceeds a cost of, say, $100,000) is treated as an IT project and undergoes the kind of scrutiny and management rigor appropriate to larger, riskier IT undertakings.

[**] For a graphical representation of the workflows and associated measures for the problem resolution and service request processes, see *The Hands-On Project Office*, http://www.crcpress.com/e_products/downloads/download.asp?cat_no=AU1991, chpt4~1~Service Delivery Workflows with Metrics~example. For a template with examples see chpt4~2~Service Level Management~example with templates.

[***] Your organization's problem ticket system will usually provide reports that may serve as the basis for setting performance levels. In the spirit of continuous improvement, choose a level of performance measure somewhere above the mean. Whatever level you choose, it must align with the IT resources devoted to delivery. Otherwise, your SLA process is bound to disappoint your customers.

Exhibit 1 SLA Problem Resolution Metrics

Severity Level	Customer Impact	Customer Response	Max. Resolution Time
Code 1 — catastrophic	Global service shutdown	Within 30 minutes	ASAP
Code 2 — urgent	Failure of a major enterprise system or network leg	Within 60 minutes	ASAP
Code 3 — high priority	Individual or team severely impacted	Within 4 hours	2 business days
Code 4 — important	Individual or team minimally impacted	Within 3 business days	As scheduled

Of course, these are averages, but in quantifying how well the IT organization actually performs against these metrics, the CRE has a clear and concise tool for communicating to his or her customer both the value of IT services as delivered and any actual performance issues. Similarly, IT can identify problems in its own performance and measure improvements once corrective actions are taken.

Similarly, your help desk process and the associated SLA document must address the topic of service request metrics. Although Exhibit 2 appears to be the same as its predecessor, the underlying activities and participants in each of the coded service request workflows are somewhat different from the problem resolution model.

Like the problem resolution matrix, the service request matrix prioritizes work based on the highest value to the greatest number, concluding with requests that almost verge on a project status but that IT has agreed to address through more routine maintenance and enhancement activities. The key to success in either of the these measurement models is adequate discussion before rolling out the service management model, to ensure that the levels, as set, are acceptable to customers. Here, the IT organization may offer a level of service based on available resources and reasonableness, allowing the business units to negotiate higher levels of service if need be (and if realistic) for increased levels of business unit contribution. In reality, from time to time IT must deliver on service requests at the eleventh hour and usually at the behest of a senior enterprise executive. Some of these requests are driven by unforeseen business opportunities or crises. Others are merely due to poor planning. Whatever the reason, will your CIO say no to the president of the enterprise in such a circumstance? Service providers should anticipate

Exhibit 2 SLA Service Request Metrics

Severity Level	Customer Impact	Customer Response	Max. Resolution Time
Code 0 — VIP request	Request is from a senior executive who needs an IT service unexpectedly and right away	Within 2 hours	ASAP
Code 1 — urgent request	Request is time-sensitive and must be completed or it will impact a large subset of the enterprise	Within 4 hours	1 business day
Code 2 — high priority	Request is time-sensitive and must be completed or it will impact an enterprise team or service	Within 1 business day	2 business days
Code 3 — standard request	Request is for some routine service in the future	Within 2 business days	5 business days
Code 4 — special request	Request requires specialized assistance and a considerable amount of work but is not time-sensitive	3 business days' notice at a minimum	As scheduled

these events and account for them in the overall SLA schema, just as the IT organization will need to accommodate such requests in its allocation of personnel and other IT resources.

No service level commitment will hold unless the IT service providers in question can validate their own ability to meet the service levels offered in the SLA. To that end, they must decompose their processes and measure process outcomes, adjusting metrics as needed and, one hopes, before presenting them to the customer for approval. Even so, the final test is in the doing. Do not be surprised if your initial foray into SLA metrification requires a midcourse correction or two. As your problem tracking system compiles data, your team will learn how long it truly takes to resolve particular problems and deliver support and maintenance services. The averages that come out of your measurement tools will let you know. Perhaps you will need to reengineer processes for efficiency or add resources because of increased demand. Sometimes, you will need to find other ways to provide needed services, such as outsourcing a particular service to a more proficient third party. Do not be afraid to admit that your IT organization is not all things to all people. Your performance measures will assist you in justifying your course of action and in building customer support for that course of action. Your organization's CREs will provide an essential role in managing the ongoing expectations of business unit customers and ensuring high quality, continuous communication among all parties in this service delivery management process. Thanks to the tools at hand and the analytical and operational support of the PMO, the CREs will have an advantage in dealing with your key customers.

Truly effective IT service organizations get that way in part due to collaboration among their service delivery units. For example, the help desk or call center must maintain an effectual intake process. To the extent possible, customer issues should be addressed then and there, rather than being passed off to another IT team. When others beyond the help desk or call center must address the customer's need, handoffs should be timely and properly directed. The receiving IT team must address customer problems and requests in keeping with the priority codes and service levels agreed to within the process. To ensure that these activities are well coordinated, IT organizations may even rely on internal SLAs to set clear, measurable performance standards among IT service units.* The author certainly endorses this approach, but it may not be realistic for enterprises in which the culture of collaboration is more informal.** Whatever form the effort takes, the IT team must agree on service levels, the timing of handoffs, and measurement and reporting standards.

In many instances, IT organizations address problem resolution and service delivery management through a three-tier model, in which the help

---

* In getting Northeastern University's IT service providers all on the same page, information services (IS) adopted performance standards and metrics as a team process and then employed staff training to educate each member of the IT staff about how his or her work would be measured according to these new standards. For an example of the associated training tool, see *The Hands-On Project Office*, http://www.crcpress.com/e_products/downloads/download.asp?cat_no=AU1991, chpt4~3~SLA~template. For an illustrated application, see chpt4~4~SLA~example. For a hardcopy version of this tool see Appendix D.

** Some IT executives worry that establishing internal SLAs will provide their people with an excuse to perform only at the levels set in the agreement, or that service personnel will hide behind an internal SLA when services fail. These concerns may be addressed as part of the process, but it will not be easy. The IT management team must first create an environment in which collaboration is rewarded and performance measurement is viewed as an essential tool for continuous process improvement. Although this is a challenging undertaking, the rewards of a successful application of the principles embodied in this process are great.

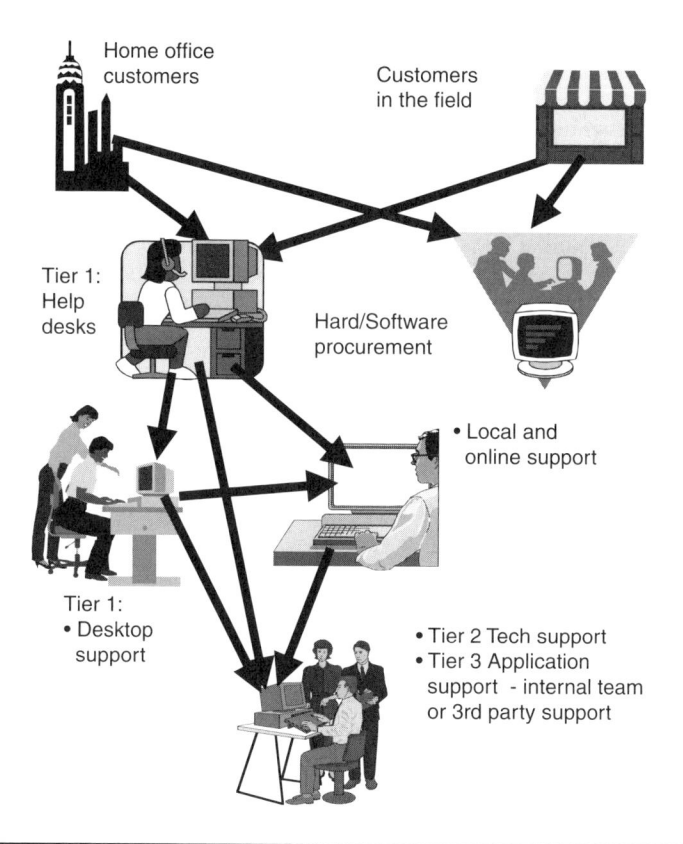

Exhibit 3 A Three-Tier Service Delivery Model

desk or call center serves as Tier 1. At the Tier 1 level, IT personnel do the entire intake process with an objective of addressing 80 percent or more of all problems over the phone. When resolution by phone is not possible, the problem or request is escalated. A field support team, also possibly operating within Tier 1,* may be dispatched to address desktop and printer issues and to install hardware and software locally. See Exhibit 3. For those problems requiring a higher level of expertise than that found within the help desk and field support teams, tickets will be forwarded to Tier 2's internal teams of specialists (e.g., network services, client server systems, and application development teams). These experts will provide the fixes required to get a particular IT product or service back online.

---

* In typical IT parlance, Tier 1 support refers to help desk or call center intake support. Tier 2 support refers to work done by the internal IT support unit responsible for that product or service. Tier 3 support refers to a service need that typically exceeds in-house expertise and thus involves both the internal IT service unit and an external technology partner provider in addressing the customer need.

When the issue suggests a larger hardware or software problem, requiring the replacement, upgrade, or reengineering of an IT product, the ticket will escalate to Tier 3, where IT vendors and other external partner providers may get involved. From the perspectives of cost and customer satisfaction, it is far better to address an issue in Tier 1 rather than to have that issue work its way to Tier 3. Invest in a skilled and highly cross-trained help desk team and build IT solutions around standardized and well supported technologies. If your service teams are meeting 80 percent of customer needs through Tier 1 services, 15 percent through Tier 2 services, and 5 percent through Tier 3 services, you are doing very well indeed, especially when the knowledge gained from these experiences is recycled across the IT organization in the spirit of continuous improvement. Employ the PMO to codify and disseminate this learning.

By measuring and reporting on performance, IT management has an advantage in dealing with overall customer satisfaction. And this is the name of the game. To ensure that IT comes away appreciated and respected by the customer, take care to ensure the quality of the Tier 1 experience. Complement this effort by making it crystal clear what your business colleagues should expect in service through an SLA process. In this manner, IT will be managing the situation from both the bottom up and the top down. Together, these processes will make it easier to work in partnership with the business side of the enterprise and to gain the recognition that your IT team deserves. Let us now consider the SLA document in more detail.

THE SERVICE DELIVERY AGREEMENT

The SLA is first and foremost a framework for discussions about the customer's service requirements and the IT organization's ability to deliver on those requirements.[*] In some enterprises, the SLA may serve only as a script, framing and organizing face-to-face discussions between IT management and its business unit counterparts. In other enterprises, the SLA constitutes a formal contract between IT and its customer. This is especially true when the business unit pays IT directly for services provided. At the very least, the SLA should offer the business unit a clear, comprehensive picture of the IT products and services it receives and the level of maintenance and support associated with each offering. The details captured in the document will reflect its anticipated use as a talking piece, a price list, or a binding business contract. Generally speaking, SLAs cover the following subjects:

---

[*] For an electronic version of the SLA template and an example, see *The Hands-On Project Office,* http://www.crcpress.com/e_products/downloads/download.asp?cat_no=AU1991, chpt4~3~SLA~template and chpt4~3~SLA~example.

- SLA process objectives
- Basic service terms and definitions
- Roles and responsibilities of process participants
- An inventory or summary listing of hardware and software assets
- Standard service process workflows for problem prioritization and resolution
- Costs of services encompassed by the SLA*
- Desired SLA process outcomes and performance metrics

Because this chapter has already considered the SLA process in some detail, the next section will address the remaining components of the SLA.

BASIC SLA TERMS AND DEFINITIONS

From the outset, your SLA document and process must make clear to the customer what services are encompassed by the agreement and what services are excluded. The following definitions may be drawn upon in constructing a customized SLA document:

- Maintenance is defined as any activity, performed at the discretion of IT, that invests in and preserves the value to the customer of an existing IT application and environment including the following:
  - Defect correction — this is the correction of critical defects found in a deployed application which inhibit the enterprise from meeting its production system availability or performance requirements. Examples of defect correction activities include responding to production calls for batch systems running overnight and installing system bug fixes.
  - Retooling — this includes any required change related to an upgrade of an infrastructure vendor product. An example would be changes needed to support CICS, DB2, etc.

---

* The costs documented in the SLA may include both nondiscretionary IT costs, such as vendor-based software licensing and maintenance fees, and the discretionary costs associated with system and Web site enhancements. In practice, each business unit's SLA funding level may be set to meet all nondiscretionary cost requirements. Discretionary expenditure funding may then be allocated based on past activity and anticipated needs. The total enterprise's IT investment in these cost categories and the distribution of the associated funds across business units would typically be set by the enterprise's planning and governance process as part of annual budget planning. For enterprises that fund the IT organization directly, actual cost information may be replaced by a list of the assets covered or perhaps a list of the IT personnel dedicated to that customer's service issues.

- Asset protection — this ensures the continued operation of any production system when changes to enterprise platforms are implemented. An example would be a DB2, ColdFusion, or CICS upgrade.
- Disaster recovery procedures — these support the business unit in developing its disaster recovery plan and participating in any disaster recovery testing.
- Required by external agencies — these are activities required by external and internal audit, regulatory agencies, etc. An example would be the process to move applications to and maintain them in a formal change management environment.
- Applied research and feasibility analysis — as part of both SLA and project work for the business unit, IT services will conduct assessments of IT products, services, and processes to determine their appropriateness in line with business unit needs.
- Infrastructure and related production support — this is work associated with the implementation and operation of business system applications, especially as these activities relate to Internet-enabled applications.
- System support — this includes tier 2 and 3 help desk and phone support for existing applications, including both break/fix support and related customer (i.e., end-user) assistance.

■ Enhancement work is defined as the day-to-day, business-as-usual activities required to satisfy customer-driven technology platform and systems application requests, including the following:
- Scheduled updates — these include regular upgrades, bug fixes, and enhancements from the external manufacturer or supplier of an application system installed on behalf of the customer's business.
- Problem management — this is the process of identifying, investigating, resolving, and preventing events or conditions that interrupt the user's ability to perform his or her job function.
- Change management — this is the process of knowing, controlling, and planning the configuration, organization, implementation, and operation of products and services. Such changes would typically reach production through regularly scheduled releases of bundled upgrades and enhancements.
- Enhancements — these are defined, customer-requested improvements or expansions to an existing application's business features or functions. The request does not have mandated start and end dates, and often falls below the priority of a development project request. The criteria for classifying a request as an enhancement are as follows:

- The request has no significant external dependencies (i.e., no web of commitment outside of IT and the business unit in question).
- The request does not exceed an estimated $100,000 in total (or whatever clip rate is set by the IT organization).

In reviewing these definitions with the customer, the CRE establishes a common vocabulary for describing and requesting IT services. The boundaries and rules of engagement placed around each service component set expectations. Typically, maintenance service definitions are no problem, and any vagueness about what constitutes an enhancement and what constitutes a project is sorted out over time and through mutual agreement.

ROLES AND RESPONSIBILITIES OF PROCESS PARTICIPANTS

Service delivery management process participants will vary from enterprise to enterprise. Here is a list of the more typical roles and their descriptions:

- Sponsor — this is the executive leader of the business unit who ultimately approves the funding for the business unit's IT work under the SLA.
- Working client — these business unit managers work alongside their IT services counterparts to define, develop, and deliver IT products and services to the business unit.
- Customer relationship executive (CRE) — this IT executive is ultimately responsible for the satisfactory delivery of the IT commitments consolidated under the SLA.
- IT finance — IT service's controller provides information during the SLA preparation process and reviews and approves all SLAs from a financial perspective to ensure that they align with the IT organization's overall commitment to the enterprise.
- IT project and service team leaders — these parties, who are directly responsible for the delivery of products and services to working clients, coordinate the efforts of their respective teams and maintain ongoing communications with the working clients as required to ensure the quality of IT deliverables.
- Project management office — as needed, the PMO assists CREs in the framing of SLAs and will partner with product and service teams in the development of the project plans, business requirements, functional specifications, and so forth.
- Enterprise corporate management — together, these individuals ensure that, as part of the annual planning and IT funding approval processes, they review and approve the overall IT SLA process framework and its associated resource allocation model.

Even if the business unit accepts these role definitions, your partners may not appreciate their responsibilities as part of an SLA. Indeed, business unit leaders often look upon the SLA as a list of things that IT must do for them, rather than a mutually obligating agreement. Therefore, include within the SLA itself an explicit list of business unit responsibilities. The CRE must review these working assumptions with the business unit team as part of the initial SLA review, and if and when any of these business unit commitments becomes an issue. Here is an illustrative list of what the business unit sponsors and working clients might commit to as part of the SLA process:

- Operating within the information technology funding allocations and funding process as defined by the enterprise's governance, planning, and budgeting processes
- Working in close collaboration with a designated IT CRE to frame this SLA and to manage within its constraints once it is approved as part of the budget for the fiscal year
- Collaborating throughout the life cycle of the project or process to ensure the ongoing clarity and delivery of business value in the outcomes of the IT effort, including direct participation in and ownership of the quality assurance acceptance process
- Reviewing, understanding, and contributing to systems documentation, including project plans and training materials, and any IT project or service team communications, such as release memos
- Throughout the life cycle of the process, evaluating and ultimately authorizing business applications to go into production
- Distributing pertinent information to all associates within the business unit who utilize the products and services addressed in this SLA
- Ensuring that business unit hardware and associated operating software meet or exceed the enterprise's minimum hardware and software standards
- Reporting problems using the reporting procedure detailed in this SLA, including a clear description of the problem
- Providing input on the quality and timeliness of service
- Prioritizing work covered under this SLA and providing any ongoing prioritization needed as additional business requirements arise
- Employing the enterprise's IT standards and architectures whenever possible and recognizing the TCO implications of failing to observe these standards

By winning the customer's acceptance of these ground rules, the SLA process creates a working relationship that fosters effective collaboration and, ultimately, the successful delivery of IT services.

REPRESENTING IT ASSETS AND COSTS

If your organization must account for its IT costs by customer group, the SLA should subdivide all IT products and services, and their associated costs, into two groupings: maintenance and enhancements. This requires collaboration among your organization's business officer, your service delivery line managers, and PMO personnel. Within these categories, the document can also differentiate between vendor-based and internal products, services, and costs. When in doubt, consult your customer about how he or she would like to see this information presented. All related costs (e.g., asset protection, default correction, license upgrades) should be grouped together for ease of customer communication and understanding. For each business application serviced through the SLA, there will be associated labor (internal and external or contract), consulting, software, and hardware costs.

Although some of these costs are generated through the consumption of internal labor resources, others may be incurred through third parties:

- IT product vendor costs may encompass all charges related to licensing or purchasing hardware and software, as well as contract labor (time and materials) or consulting (fixed price) costs for product implementation or subsequent servicing.
- To ensure that the best and most appropriate resources are deployed to meet business unit requirements, IT may regularly employ external contract labor, billed on a time-and-materials basis, to complement in-house staffing resources.
- When the IT organization lacks a strong internal task or talent match in undertaking certain types of IT work, when there is a need to ramp up an effort quickly, or when IT has lost a key internal expert, IT will regularly employ external professionals in specialized areas on a consulting basis, typically billed to the customer on a fixed-price contract basis.

A simple template suffices to capture and communicate this detail to the customer. See Exhibit 4.

Enhancement costs are distinct from those for maintenance in that they are largely discretionary. They are defined as customer-requested changes to existing applications. Typically, the IT organization will negotiate an allowance for business unit–enhancement work. More often than not, these requests do not have mandated start and end dates, and they often fall below the priority of project requests. An example of the criteria for classifying a request as an enhancement follows:

- Request has no significant external dependencies
- Request is below a certain clip rate, such as $10,000 per instance

Exhibit 4   Maintenance Cost Matrix

Business Application	Labor Costs	Consulting Costs	Software Costs	Hardware Costs	Total Cost
Enter the name of the appropriate business application	Enter internal and external labor costs	Enter associated consulting costs	Enter application-specific software costs	Enter application-specific hardware costs	Add columns 2–4

Although it merits its own section to record tasks and costs, the enhancement cost matrix is otherwise identical to the maintenance cost matrix. See Exhibit 5.

A word of caution: take your time and be accurate. Nothing destroys trust within an SLA negotiation more quickly than misrepresenting or misunderstanding the underlying costs of the services governed by the agreement. So do your homework before meeting with your customer. Learn how the customer consumes your services, as well as how you intend to deliver them. You also must enter the conversation with a firm grasp of your business' cost drivers. If the customer responds to the SLA by saying, "We need to cut costs by 20 percent," you must be prepared to explore such issues to ensure the most positive outcome for the customer. Last but not least, be sure to state explicitly those services excluded from the scope of the SLA, in particular large-scale enhancement efforts that are estimated to exceed the SLA clip rate. These should be treated as separate IT projects. Do not leave the table

Exhibit 5   Enhancement Cost Matrix

Business Application	Labor Costs	Consulting Costs	Software Costs	Hardware Costs	Total Cost
Enter the name of the appropriate business application	Enter internal and external labor costs	Enter associated consulting costs	Enter application-specific software costs	Enter application-specific hardware costs	Add columns 2–4

until you have reached consensus and your business partners have signed the SLA document.

PROBLEM RESOLUTION AND SERVICE DELIVERY WORKFLOWS

As a convenience to the customer and as another expectation setter, many SLAs identify the names, roles, and even phone numbers of appropriate IT support personnel. These are more granular operational issues than those of the overall cost and scope of service delivery, but they belong in the SLA all the same. For example, the document might also include standard hours of operation and coverage for IT services, such as network operations, system support, the help desk, and so forth. Sometimes, it may prove useful to include the schedule for production services report runs and system refreshes. All this must be stated succinctly and in nontechnical language. Where practical, tailor the presentation to the needs of the particular business unit.

Finally, the SLA should include the service level metrics discussed previously. This information makes it clear to the business unit that IT will respond to and address its needs based on business-driven priorities and that all customers will be treated in an equitable manner (except VIPs, of course). You may also decide to include the actual workflow for problem resolution, to clarify roles, responsibilities, and the anticipated timing of deliverables. A simple example of this approach follows:

- Priority 1
  - Definition — the application is unavailable to anyone at an enterprise work site.
  - Response time — work will begin immediately and continue until resolved.
  - Responsibilities
    - IT CRE — resolves problem and communicates with all who are affected at least daily until resolved.
    - Working client — works alongside CRE until the matter is resolved.
    - Partner providers — other IT teams and third parties, as appropriate, provide technical assistance.
- Priority 2
  - Definition — the application is not available for individual users within a site.
  - Response time — a response will be provided within one business day. A recommended solution will be provided within three business days if there are *no* outstanding priority 1 problems. Finding

a solution to a priority 2 problem will not begin until resolution of all priority 1 problems that impact the priority 2 issue's resolution.
- Responsibilities
  - IT CRE — sends acknowledgment of problem; resolves problem and communicates status to all who are affected.
  - Working client — works alongside CRE until the matter is resolved.
  - Partner providers — other IT teams and third parties as, appropriate, provide technical assistance.

■ Priority 3
- Definition — the application generates appropriate results but does not operate optimally.
- Response time — improvements are addressed as part of the next scheduled release.
- Responsibilities
  - IT CRE — communicates needed changes.
  - Other process participants — commit to work as part of the regular system upgrade cycle.

REPORTING ON RESULTS

The keys to successful IT service delivery are, on the one hand, explicit standards and measures of performance, and on the other hand, the consistent and uniform reporting of performance results. Each side of this coin presents its own challenges. In terms of standards and measures, all too often IT personnel track activity rather than results. Typically they measure what was done rather than how closely these activities meet the customer's needs. For example, a field service organization might report on the number of desktops that it has installed in a given month. This statistic is a measure of activity. It does not tell us, however, if any of these implementations were done in keeping with customer timetables or if the associated software installs worked afterwards, or if the technician properly trained the customer to make best use of the system before leaving the customer site. The answers to these questions are actual results measures.

Clearly, activity counts are important in reporting on the scope and volume of activity, but they do not tell the whole story. At the same time, the primary focus of the IT organization is actual product and service delivery, not reporting on it. In devising metrics, ask the following question: For any given IT service, what drives customer satisfaction, and how do I best measure this? At times, you will be obliged to settle for a surrogate metric because the true measure of customer satisfaction is too difficult to ascertain. By way of illustration, here are some of the metrics that the author has employed in measuring the performance of his IT team:

- Network services
  - Number of problem tickets logged, closed, and remaining open in a given month
  - Network availability (by segment or location if appropriate)
  - Network response time (by segment or location if appropriate)
- Call center
  - Calls received
  - Calls abandoned
  - Calls addressed without escalation
  - Number of problem tickets logged, closed, and remaining open in a given month
- IT training
  - Number of courses offered
  - Number of seats available
  - Number of students who complete courses (i.e., stay to the end or receive certification in the specific subject)

All of these metrics allow for the ready capture of easily quantifiable performance measures. In the case of network services, overall availability and response time are what customers care about most. However, measures of response time can be deceiving. Although the network as a whole may be up and running, a particular customer segment of service (i.e., business application) may be down. Thus, you must be clear about what you are measuring and how you are measuring it. Monitoring the flow of problem tickets is one of those surrogate metrics. If your network services unit has few to no problem tickets open, this is a good thing. Similarly, a spike or growing backlog in problem tickets would suggest that something is amiss. For the call center, which is responsible for the intake of customer problems, measures of efficiency in call handling and ticket processing are also important. By the same token, growth in abandoned calls would suggest issues with the intake process. In some instances, you will have more concrete measures of service delivery effectiveness. For example, if you test students on their knowledge following an IT training session, the results would serve as a clear indicator of instruction and course effectiveness.

Unfortunately, none of these metrics truly suffices because none indicates the satisfaction of the customer. To measure customer satisfaction, you must poll customers, asking them simple questions about what they liked or did not like about their interactions with IT. Here too, you must focus on those areas that get at what your customer expects from the service. For example, in surveying a recipient of audio/visual services, you might ask the following questions:

- Was the equipment delivered in a timely fashion?
- Was the equipment in proper working order?
- Were the A/V personnel knowledgeable about equipment operation and uses?
- Did the staff provide the user with sufficient instruction so that the user could operate the equipment on his or her own?

Similarly, in the case of desktop support, you could ask questions like these:

- Was the support timely?
- Did it address the need?
- Did the IT employee seem knowledgeable?
- What was the customer's overall satisfaction with the quality of the deliverable?

In each of these examples, a few questions get to the heart of the service instance and establish the customer's satisfaction with the experience.*

Once you have designed your survey tool, you must consider how best to deploy it. Your approach will depend on your corporate culture. Enterprises accustomed to internal measurement will be more receptive to survey forms and questionnaires than those where the practice is brand new. For other organizations, the only means of obtaining feedback is to call and or visit with customers one on one. If you have an enterprise intranet site, you might compromise between these two extremes by employing some user-friendly survey tool, such as eSurveyor, to poll people online. Whatever mechanism you choose, the process must be focused, quick, and painless. To begin, draw your survey population from your trouble ticket management system (i.e., survey those who have actually received service). Next, tailor the query to the particulars of the deliverable (i.e., ask questions that pertain to the specific service delivered). If the responses are unfavorable, ask for details or examples and whether an appropriate service manager may follow up with the customer. Track and consolidate your results to get an overall sense of IT team performance. The mere act of asking the customer is a win-win situation for the enterprise. On the one hand, it keeps the IT team customer-focused, better aware of customer needs and issues. On the other hand, the obvious effort at continuous improvement wins IT friends and support

---

* For a complete example of how the various metrics for IT service delivery may be captured and reflected in a single management tool, see *The Hands-On Project Office*, http://www.crcpress.com/e_products/downloads/download.asp?cat_no=AU1991, chpt4~5~customer satisfaction measures~example.

on the business side of the house. Indeed, the typical customer response runs something like this: "They actually care!"

The process does not end with data collection, however. Once collected, data should be aggregated, analyzed, and shared across IT, and perhaps with customers as well. My preferred mechanism for doing just that is an operations report. This is a monthly activity for which the IT management team comes together to scrutinize both SLA performance and project delivery. Because there tends to be a lot to talk about, you might consider dividing your own process into two segments: one for service performance and the other for project statuses.* The operations review and report focus the IT management team on customer service. Organizing and running the monthly sessions, as well as creating the report itself, fall to the PMO. The service review session should include all those with line responsibility for service level management; the project review should include senior IT managers and all project directors. The remainder of this chapter will consider the SLA side of the operations report, leaving the project side for detailed consideration in Chapter 5.

Meeting the standards set in the IT organization's SLAs is a communal responsibility. Therefore, successes and failures both should be explored collectively in the hope of identifying and addressing the root causes of service delivery problems. The focus is not on blame but on self-improvement. By participating in constructive inquiries into these matters, IT management demonstrates its commitment to a customer focus and to the team's continuous improvement. In larger and geographically dispersed IT organizations, the monthly meetings are also an opportunity to get all the players in the IT value chain around the table to share information and to coordinate assignments that cut across silos. For example, if data center management plans a shutdown on an upcoming weekend, the operations review session affords an opportunity to remind colleagues and to identify any issues that might arise from the planned event.

On a more interpersonal level, the prospect of a monthly review process encourages team members to get and keep their respective houses in order. No matter how friendly the atmosphere of the meeting might be, no one wants to parade bad news in front of one's peers on a regular basis. Rather than allow things to fester or, worse, to explode at the session itself, participants will sort out their issues with colleagues

---

* For an "Operations Report A" template, see *The Hands-On Project Office*, http://www.crcpress.com/e_products/downloads/download.asp?cat_no=AU1991, chpt4~6~monthly service delivery report~template. For an example of a completed template, see chpt4~7~monthly service delivery report~example.

offline and bring forward solutions, not added conflict. If IT management sets the right tone at these sessions, over time they will have a constructive impact on the culture of the organization, break down the siloed mentality among IT departments, and promote collaboration. Thus, what may at first appear to some as added administrative overhead will go a long way toward building a better appreciation of the challenges and opportunities faced by IT in consistently delivering high-quality services to customers. The actual operation of these processes falls to the IT organization's PMO function. Working with the CREs, PMO personnel will ensure that the activities and documentation described previously are well coordinated and delivered to IT management and IT's customers in a timely fashion.

The service delivery components of the operations report include both qualitative and quantitative customer service measures. Each service unit should report on outages and other events that impact IT's customers. Your report should require a clear measure of the impact (e.g., the duration of the service disruption and the audience impacted) as well as the steps taken by the IT organization to address the problem. At the operations review meeting, this information will serve as a basis for discussion on why such problems occur and how to prevent them from recurring. The report will also include quantitative data drawn from the problem ticket system, customer surveys, and other measures reflecting this month's IT team performance against SLA metrics. Here, too, the point of the meeting is to compare the current month's performance against past months and to speculate on how best to improve service performance. Such discussions are good for the team and ultimately for the customer. Make this continuous improvement process part of the way you do things.

The report findings will prove useful to the CREs as they meet with their customers and communicate the extent and quality of the IT organization's service commitments to each line of business. If the report identifies problems, the customer will already know about these but will be comforted that IT is also aware and taking steps to correct the situation. Finally, as a body of information, the unit's operations reports serve as an excellent chronicle of day-to-day IT team performance. The PMO knowledge management process can employ this data to track improvements over time and identify areas in need of greater management focus or technology investment. The results of best practice are captured for all to see, with the expectation that service teams will learn from one another. Lastly, when next year's budgeting and planning process gets under way, the CIO will have easy access to information that demonstrates which IT investments paid the greatest return over the past year.

CLOSING COMMENTS

This concludes our overview of IT service delivery management. The critical success factors associated with this complex set of processes may be summarized as follows:

- A common agreement among all participants, including line-of-business customers and IT service providers, about roles, responsibilities, and principles of operation
- Clear, well communicated rules of the road, workflows, and metrics that set customer expectations and measure IT accountability
- Ongoing process measurement and improvement with a focus on customer needs and on appreciating the value of IT to the enterprise
- Regular reporting through informal and formal communication venues, both within the IT organization and between IT and its customers
- Proactive listening on the part of the CRE in dealing with sponsors and working clients and in interacting with his or her IT colleagues
- The employment of the PMO to
  - Codify metrics
  - Collect and analyze customer feedback
  - Prepare SLA documents
  - Prepare the monthly operations report
  - Coordinate monthly operations report meetings
  - Provide staff support to the CRE function
  - Maintain the archives of performance measures, SLAs, and operations reports
  - Oversee and continuously improve all the processes that underpin IT service delivery management
- Collaboration all around and attention to the details

By adapting the frameworks and models presented in this chapter, your IT organization can excel at routine service delivery. In so doing, you will earn the credibility and support that you must have to win approval for the enterprise's further investment in IT platforms and systems. When IT takes care of the day-to-day and effectively manages customer expectations, the business community will grow more tolerant of the frustrations and high costs associated with the implementation of any large-scale, complex IT undertaking. At the very least, your customers will better appreciate the value of existing IT services and the necessity of partnering to enable the success of future efforts. To that end, use the tools presented here and on the accompanying Web site (http://www.crc-press.com/e_products/downloads/download.asp?cat_no=AU1991) at your own discretion to create and maintain a healthy SLA process. Feel free to adapt my examples so that they apply to your own workplace.

Having considered the bread-and-butter issues of service delivery in some detail, it is now time to move on to a more detailed consideration of IT project management. Here, too, the PMO has a major — if not central — role to play.

5

PROJECT DELIVERY AND THE PROJECT MANAGEMENT LIFE CYCLE

INTRODUCTION

Clearly, the ability to manage and deliver information technology projects is key to the success of any IT organization. Although project delivery may take second place to the time, effort, and resources devoted to service delivery, the IT team's performance on projects, in many respects, defines IT's overall status within the enterprise. If projects are managed well and delivered as promised, the IT organization builds credibility and hence the support and resources to take on an expanded role within the business. If it falters in these efforts, it loses the credibility and political capital essential to enabling a strategic role for IT within the enterprise. In short, if IT management aspires to a true partnership with the business side of the house, they must demonstrate serious competence in managing large, complex, impactful assignments.

Projects succeed when they are well aligned with enterprise goals, objective, and well managed. The underlying focus of project management as a process is all about managing risk. Indeed, when one considers all of the processes associated with project management, including scope and commitment management, project planning and budgeting, business requirements gathering, issue tracking, change control, quality assurance, and release management, one sees that these efforts are about limiting the risks of project failure. These precautions are necessary because, unlike routine, predictable, and ongoing service delivery tasks, IT projects encompass work that is unique, sometimes unpredictable, and time boxed. By mitigating risk and introducing rigor to the process, project management strives to make projects more assured and consistent in their outcomes.

Project success is easily measured. Project delivery must occur on schedule, within budget, and in keeping with the customer's requirements and expectations. This sounds marvelously simple. Unfortunately, the reality for many IT projects is not as simple or as happy a story. It is difficult for even the best and brightest IT teams to satisfy their customers consistently in this regard. Does this mean that the project management process has failed us or that IT teams are inherently deficient? I doubt it. But past performance does suggest that most IT organizations face serious challenges in delivering projects satisfactorily.

When problems arise, business colleagues may be befuddled by seeming confusion within the ranks of their IT colleagues. Given all the project work undertaken by the typical IT organization, they assume that IT managers have in hand a formula for repeatable success in this discipline. Little do they realize that there is no philosopher's stone available to help IT professionals through this dilemma. On the other hand, there are enough successful IT implementations to suggest that some of our colleagues have developed repeatable processes to limit project risk and to obtain favorable outcomes. Those IT organizations that do stand out have one thing in common: a rigorous and disciplined methodology for scoping, planning, managing, and reporting on their IT projects. Although many other organizations face the same challenges, few have had the opportunity to invest in the development and documentation of a proven project management tool set. Typically, they have also lacked a center of excellence, such as a PMO, that might refine their processes and ensure overall team compliance with the best industry practices.

To aid these colleagues, one can point to such excellent institutions as the Project Management Institute,[*] that are dedicated to the development of certified project management (for IT and other disciplines) and a veritable plethora of printed self-help material.[**] Much of this work is well conceived and highly focused, addressing the entire project management life cycle or its key aspects. All too often, neglect of some component of

---

[*] See the PMI Web site for details concerning training programs and institute publications: http://www.pmi.org.

[**] The literature on project management is vast. Here are a few recent works of interest: Project Management Institute, *A Guide to the Project Management Body of Knowledge, 2000 Edition* (Newtown Square, PA: PMI, 2000); Ron Black, *The Complete Idiot's Guide to Project Management with Microsoft Project 2000* (New York: QUE Press, 2000); Joseph Philips, *IT Project Management: On Track from Start to Finish* (New York: McGraw Hill/Osborne Media, 2002); Richard Murch, *Project Management Best Practices for IT Professionals* (Upper Saddle River, NJ: Prentice Hall, 2000). See the Selected Readings section in Appendix K for additional recommendations.

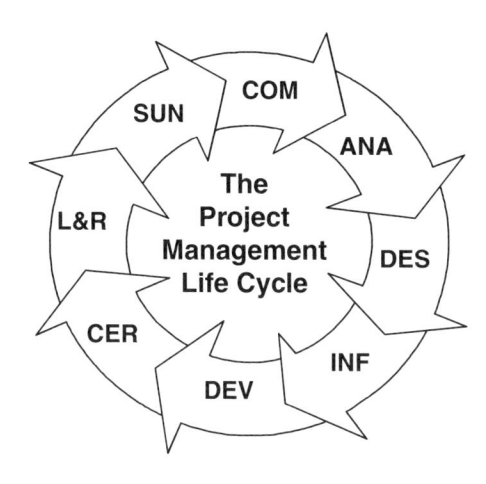

Exhibit 1 The Project Management Life Cycle. (For an electronic version of the development life cycle graphic see chpt5~/~project management life cycle~graphic. For a hardcopy version of this tool see Appendix E.)

project life cycle delivery spells doom for the effort. The PMO champions adherence to such a disciplined methodology, sometimes referred to as a project engineering framework. Exhibit 1 succinctly captures the dimensions of this formal approach.

This graphic identifies all the stages of project management, including commitment (COM), analysis (ANA), design (DES), infrastructure readiness (INF), development (DEV), certification (CER), launch and release (L&R), and sunset (SUN). It also depicts the process as a series of life cycle events, which aptly conveys the nature of the undertaking — from gestation through birth, development, launch, and finally to sunset. Many books address the project management life cycle as a whole, consider some aspect of it (like design, development or testing), or look at project management from a particular angle, such as implementing a Web-services or an enterprise planning resource system solution. The life cycle model will serve as the organizing principle for the bulk of this chapter.

It is beyond the scope of this book to summarize the vast array of viewpoints advocating good project management practices. Instead, I offer a set of experiences that collectively afford the reader a framework and methodology for thinking about project management in its various dimensions. My objective is to allow the reader to reflect on my practices and to compare them with his or her own. Adaptable tools and techniques will set the reader on the road of self-examination and continuous improvement. As you discover the particular needs and weaknesses of your own IT organization, you may then turn to more specialized texts and advice to fill any gaps in my narrative.

This chapter introduces a comprehensive framework for the design and delivery of IT projects. Beyond this model, special attention is paid to two of the most common pitfalls in IT project execution:

- Initial project scoping
- Ongoing customer expectation management and project reporting

When IT projects come up short, more often than not these failures are due either to inadequate up-front scoping and commitment management by the project team or to serious miscommunication among stakeholders throughout the project life cycle. As always in this book, this chapter is built upon field-tested methods and tools. Whatever the scope and nature of the reader's own project management woes, these pages offer a series of applied examples that will prove useful in the self-assessment of in-house IT team practices. Adapt these to the particular requirements of your business.

WHAT IS AN IT PROJECT? WHAT IS PROJECT MANAGEMENT? WHY BOTHER?

Most IT organizations provide services in any one of three logical categories. First, IT solves customer problems in support of existing products and services. This type of service usually involves a help desk or call center, hardware and software support personnel, and training and documentation services. The objective of problem resolution is to address the end user's specific performance issues as quickly and as painlessly as possible. Second, IT responds to service requests that call either for the extension of existing products and services to a new employee or for the modest expansion and enhancement of an established product or service to existing employees. Here, too, a help desk or call center is often the intake mechanism for service request work, typically complemented by dedicated support and maintenance teams assigned to customer servicing and delivery. Neither problem resolution nor service request efforts entail large capital outlays or major changes in platform technologies. In most instances, however, they do consume significant IT team resources to service, fix, or build upon what is already there. Generally speaking, the customer's expectation is that delivery will be immediate or nearly so.

The third category of IT team activity — projects — encompasses the significant expansion of existing products and services or the introduction of new ones. Unlike the first two categories, project delivery typically requires major capital outlays; larger, more complex scopes of work; a project management infrastructure; the involvement of external technology partner providers; and long (as opposed to short or immediate) delivery timeframes. To differentiate projects from maintenance and support work, most IT organizations establish a quantifiable threshold level for the project designation and its associated management overhead. For example, an IT undertaking is typically viewed as a project

- If the IT investment under discussion leads to the establishment of a new IT product or service
- If the IT investment exceeds a certain dollar amount, where the clip level for a particular project varies from enterprise to enterprise (e.g., $10,000 for some; $100,000 for others)
- If the IT investment requires the participation of two or more IT operating units or external partner providers for delivery (i.e., scope and complexity factors)
- If the IT investment involves high levels of complexity or risk (e.g., multiple technologies, emerging technologies, systems integration)

Bear in mind that the entire purpose of project management is effective risk mitigation so as to deliver IT projects on time, within budget, and in keeping with customer expectations. New IT products and services typically require broad-based participation and call for business process, as well as technological, change. These efforts will benefit from the rigor of a project management methodology that comprehensively pursues IT delivery requirements and process workflow adaptation. Similarly, if the parent organization is investing a significant sum of money in an IT project, the executive team will expect a professional approach that ensures the realization of its business objectives and a reasonable return on the investment. Furthermore, when many players, especially across business-unit silos, complex work processes, and diverse or new IT components are called upon to deliver the IT solution, the challenges to successful delivery are high. Here, too, it makes sense to rely on the discipline of systematic project management practices.

Once the IT organization designs and implements a common process in keeping with its own needs and corporate culture, project management will help streamline delivery and eliminate more costly one-off exercises. The key to success here is the broad adoption of simple, uniform — yet flexible and well documented — processes by IT personnel. Flexibility is important because each project and each customer relationship has its own unique dynamics. With a project in which the IT organization and the customer have lots of experience together, the management process can be more informal. Conversely, high-risk projects demand considerable rigor in the application of project management principles. But do not confuse rigor with rigidity. The process must be pliable, refocusing work effort and related resources as business requirements evolve and as emerging technologies transform the IT landscape. Bear in mind this need for flexibility and adaptability when applying the frameworks discussed below.

THE IT PROJECT MANAGEMENT LIFE CYCLE — A BRIEF OVERVIEW

All projects evolve through a natural life cycle, from inception and definition to design and construction to delivery and ongoing support. The key to successful project management is to balance the need for control against the degree of risk evidenced in the project's scope. Go for a simple management framework or checklist and rely on published sources to supplement your efforts and to fill in any process gaps when confronting unusual or extremely complex projects.[*] As a starting point, consider the model shown in Exhibit 2. This framework assists the IT team in building an appropriately detailed plan and oversight process to deliver a given project successfully. Remember to adapt this tool to meet the particular needs of your project and organization. See Exhibit 2.

This framework begins with a reiteration of the various phases of the project management life cycle, including commitment, analysis, design (from both business and technical perspectives), infrastructure readiness, development, certification, launch, release, and sunset.[**] For each phase I then provide a summary list of activities, deliverables or desired outcomes, and the parties responsible. Many components of this model will be familiar; others may require a more detailed explanation. For example, the commitment phase, which is often given short shrift in actual project execution, is dealt with in some detail in the next section of this chapter. This chapter also devotes considerable attention to ongoing delivery management, communication, measurement, and reporting. To begin, let us consider each phase of the framework in a little more detail. See Exhibit 3.

---

[*] In his career, the author has created more than a few voluminous project engineering frameworks, hoping that his project delivery teams would use them when fashioning project plans and as checklists to ensure that nothing is missed during the development and testing processes. Unfortunately, these tools have died of their own weight. No one willingly refers to them, which is behind the more streamlined approach detailed in this text.

[**] This framework also serves as the starting point and checklist for creating a more detailed project plan. For a more detailed electronic version of the framework, see *The Hands-On Project Office*, http://www.crcpress.com/e_products/downloads/download.asp?cat_no=AU1991, chpt5~2~project phases~model. For illustrative workflows of the project management process, see chpt5~3~project management process~work flows and chpt5~4~project management process~example with work flows.

Exhibit 2 A Project Management Framework — Part 1

Project Phase	Project Activities	Associated Deliverables	Responsible Parties
Commitment	Obtaining sponsorship, scoping, planning, budgeting, obtaining commitment, defining roles and responsibilities, kicking off the project	Commitment doc Project plan/budget Scorecard; metrics Change process	Project director Project manager Project manager Project manager
Analysis, Part 1	Detailed business requirements gathering; initial involvement of the QA and IT customer services teams	Project coordination Business requirements Functional requirements	Project director or project manager Bus. Analyst Bus. Analyst
Analysis, Part 2	Detailed technical and infrastructure requirements gathering; initial infrastructure team involvement	Project coordination Technical requirements Technical specification	Project director or project manager IT architect Other IT partner providers, especially infrastructure personnel
Design, Part 1	Solution definition and design	Detailed functional specification Detailed technical specification Initial QA scripting CS metrics	Project team Project team QA Customer services
Design, Part 2	System prototype (if appropriate)	Nonworking prototype Focus group(s) input	Project team and partner providers

Exhibit 3 A Project Management Framework — Part 2

Project Phase	Project Activities	Associated Deliverables	Responsible Parties
Infrastructure Readiness	Finalize requirements for development, test and production environments; order hardware and software as needed	Development environment Test environment requirements Preliminary production system requirements Infrastructure SLA	Project team and partner providers Project team and QA Project team and partner providers Project team and customer services
Development	Develop solution, install application(s), build database(s), develop reporting, design process changes, build test plans, develop training, marketing, and launch strategies	Application/system(s) User/technical documentation Reporting package Training package Testing and delivery acceptance processes	Project team and partner providers Customer services Project team and QA
Certification	QA testing and release management	Scripts Testing Bug fixes Formal release strategy	QA QA Project team and partner providers QA
Launch	Regression testing, formal certification, initial piloting of solution	Pilot process Regression testing SLAs Go-live decision	Project team Project team and QA Customer services Customer
Release	Go live	System in production Customer signoff	Partner providers Customer
Sunset	Application retirement	Products withdrawn	Partner providers

Project Delivery and the Project Management Life Cycle ■ 123

Commitment

Initial scoping discussion with sponsor/working clients; OK to proceed. → Draft commitment doc. → Draft project plan → Stakeholder review and approval → Formal project kickoff by core project team

Initiate score card process

Draft project budget

Design / **Analysis**

Design solution ← Document technical requirements ← Document functional requirements ← Document business requirements

Conduct functional and technical reviews with stakeholders → If appropriate, develop prototype of desired solution → *Proceed to infrastructure phase of process*

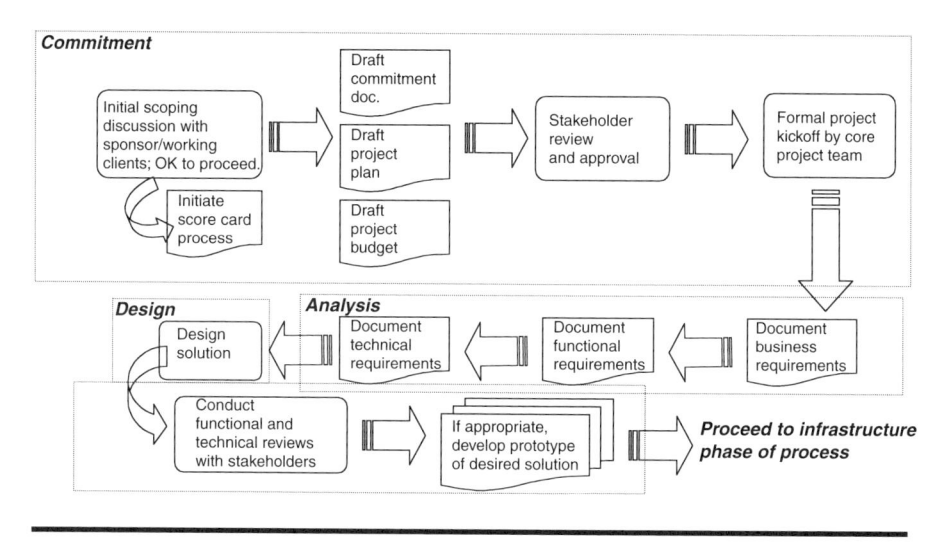

Exhibit 4   The Project Management Process — Part 1

All projects begin with some level of commitment-gathering whereby the stakeholders collectively define the scope of the effort, timeframes for delivery, resources to be allocated, and so forth. For these key players to commit to the undertaking, the commitment phase must include a preliminary roster of personnel, project plans, and budgets. Once consensus is reached on all of these key elements, the project may formally kick off and proceed to the design phase. See Exhibit 4. Depending upon the size of the project, its inherent risk, and the culture and business practices of your organization, the commitment phase may conclude with a formal, written agreement on how to proceed, or the project plan itself will constitute the document of understanding. Again, the details of this particular process phase will be dealt with in detail later in this chapter.

With commitment in hand, the now-empowered project team may proceed with its work. The analysis phase encompasses three critical elements: identifying and documenting the project's business, functional, and technical requirements. Business requirements include a detailed consideration of how the delivered IT solution will address the business sponsor's needs. As such, this requirements document should be sufficiently detailed to capture systematically all of the features and functionality that the customer expects from the envisioned system, as well as business process changes enabled by or required for its implementation. The functional requirements document translates the needs of the business unit into more IT-specific content. For example, the business requirements document may describe how an inventory control system will operate as part of enterprise operations, whereas the functional specifications define the specific functionality of the IT systems required for the desired business outcome. Lastly, technical requirements provide direction to the IT team about related hardware and software performance. Here, the document might address the volumes of data to be moved around the network, storage and backup requirements, operating system and computer platform limitations, and so forth.

Taken together, the business, functional, and technical requirements documents establish the foundation for the direction of all subsequent project activity. Note that my process model calls for both the quality assurance (QA) and the IT customer services teams to get involved in the project during the analysis phase. QA must participate at this stage in the process to better appreciate how its team should approach testing and release management when the time comes and to ensure that the project plan allows sufficient time and resources for QA activities. Customer services should also have advanced involvement in the scoping of upcoming project deliverables to prepare for IT staff and end-user documentation and training and to ready the help desk for customer support issues associated with rollout of the project.* During the analysis phase, or for that matter any other project management phase, team members may discover something that causes them to rethink the scope of the project or their approach to delivery. When this occurs, they must call together the impacted stakeholders and reconfirm their mutual commitment in light of these revelations and changes.

If the project analysis phase creates the boundaries within which you must deliver your IT project, project design defines the detailed approach and the ultimate outcome. When the project involves the installation of new hardware or software, design efforts may involve mapping the out-of-the-box functionality of these products against the customer's requirements and then addressing the gaps between that desired state and what the products in question actually offer. In this example, design work may also entail defining how best to integrate the proposed IT product into the sponsoring business unit's work processes. As suggested by the framework, more complex projects, involving business process transfor-

---

* As part of their preparation work, IT customer services and the IT unit's CREs should team up on the implications of the new project on existing service level agreements (SLAs). The framework differentiates between SLAs, which define the terms and conditions of service to actual IT organization customers, and operational service delivery agreements (OSDAs), which define the handoffs and the associated metrics among IT operating units. The purpose here is to separate customer commitments from the internal steps (and commitments!) that the IT organization must embrace to deliver on its commitments to the customer. For more details on the SLA process, see Chapter 4.

mation and the integration of various information systems, might be best served by prototyping the solution before proceeding.

Prototypes simulate the ultimate IT experience without incurring the cost, lost time, and risks associated with a full-blown development effort. In the world of Web services deployment, prototyping is a wise approach, especially for first-timers. Similarly, when you are dealing with a customer who has difficulties in articulating and settling on a specific course of action, creating a prototype may be the only way to focus the discussion and win the approval to proceed. At the end of design, the team will have its construction documents. These must be reviewed and approved by project stakeholders before any more work begins. With agreement and closure, the project may now proceed concurrently on a number of different fronts. The team will initiate the development process; it will engage colleagues if there are any networking or other infrastructure considerations; and it will work in partnership with the QA and customer services teams to ensure that these teams are ready to enable project launch and subsequent delivery. See Exhibit 5.

With the conclusion of the design phase, the project team will have firm technical requirements. This information must be shared immediately with those responsible for the IT infrastructure. I include this infrastructure phase to the project life cycle for the very reason that it is regularly neglected. Its primary purpose is to ensure that the IT infrastructure team has an early opportunity to confirm that it can accommodate, from both a server/storage and a network services perspective, any new systems or

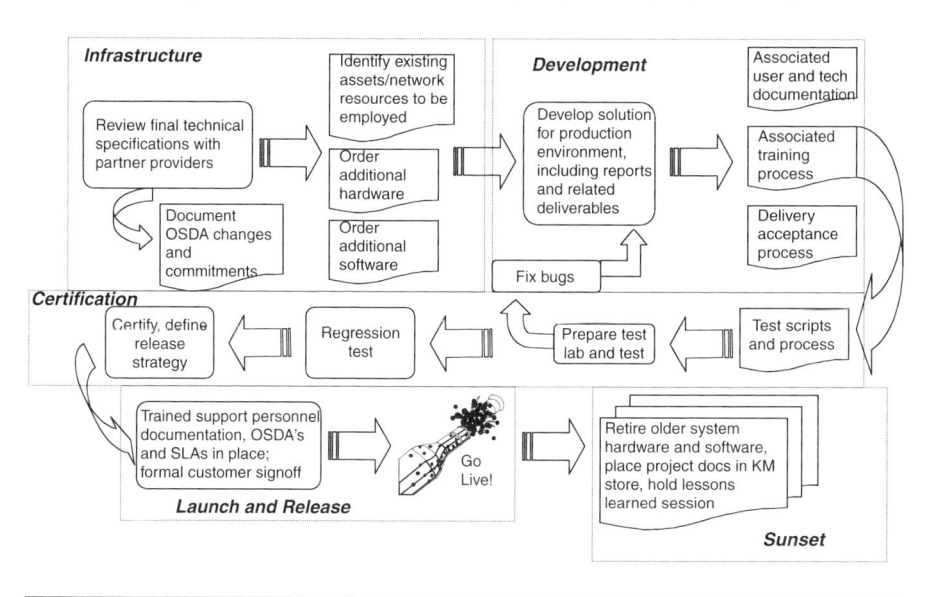

Exhibit 5 The Project Management Process — Part 2

services that your project will add to the existing enterprise IT complex. Furthermore, this phase reminds the project team to order required hardware and software early as part of project execution. It is amazing how often project teams neglect to order such things as servers before project delivery. Do not assume that your enterprise IT environment can readily absorb your new project's requirements. Even when it can, the infrastructure team must prepare for these additions and will greatly appreciate the advance notice. Finally, the infrastructure phase checklist serves to remind the project team to reserve development, test, and production platform space as required before these resources are needed.

The development and certification phases of the project management life cycle are perhaps the best documented process components.[*] Obviously, your team's approach to development will depend on the nature of the project. Make no assumptions about the level of effort required to bring a particular IT solution to market. Unless you have lots of experience with very similar projects, invest heavily in analysis and design, including prototyping, before committing to a delivery date. Remember too that there is an iterative cycle in all development efforts. As your team proceeds, it will uncover problems with the source code or the hardware or the original business and functional specifications that may require the rethinking of your development approach and perhaps the resizing of the project. You cannot always foresee such obstacles, but once they are uncovered you must address their implications before proceeding.

Similarly, do not underestimate the need for, and hence the resource demands of, the testing process. Few project plans allot sufficient time and resources for this essential development life-cycle component. Most experts swear by the benchmark that it takes as much time to test and certify an application as it takes to build it. But what project plan in your memory has ever allowed for a QA commitment to that extent? Obviously, the team must balance its investment in testing against the potential risks. Tradeoffs will inevitably occur. Nevertheless, QA takes time, especially for

---

[*] Recommended readings include: Bass, L.., Clements, P. and Kazman, R. *Software Architecture in Practice.* Reading, MA: Addison-Wesley, 1999; Freidlein, Ashley. *Web Project Management.* San Francisco: Morgan Kaufmann Publishers, 2001; Kesner, Richard M. *Information Systems: A Strategic Approach to Planning and Implementation.* Chicago: American Library Association, 1988; McConnell, Steve. *Software Project Survival Guide;* Richmond, WA: Microsoft Press, 1997; Murch, Richard. *Project Management Best Practices for IT Professionals.* Upper Saddle River, NJ: Prentice Hall, 2000; Philips, J. *IT Project Management: On Track from Start to Finish.* New York: McGraw Hill/Osborne Media, 2002; and Project Management Institute. *A Guide to the Project Management Body of Knowledge, 2000 Edition.* Newtown Square, PA, PMI, 2000.

IT applications that require extensive end-user testing, as do bug fixes and the regression testing that ensures those fixes actually solve the problem. My advice is to work with your QA team early (during the design phase) to script testing scenarios and factor the associated time and resource implications of their work into your plans.

Of the remaining aspects of the project life-cycle model, product/service release should not occur until your sponsor and working clients have reviewed and approved your project deliverables. During the course of the project, the effort's agreed-upon outcomes will change with circumstances. Be sure to reach consensus before the release of the team's work, or it may come back to haunt you later. In addition, make sure that both the IT organization and the customer are ready for delivery. Preparedness may involve operational and technical deliverables, such as ensuring that the project's production code is under version control and that system data is backed up on a regular basis, as well as support deliverables, such as the availability of technical and end-user documentation, well trained help desk and support personnel, and appropriate problem resolution and service request criteria in your call center problem management system. On the customer side of the house, pre-launch work may entail end-user training, business process (change) documentation, and production walk-throughs involving both the customer and IT system support and maintenance personnel. Be sure to publicize your launch plans well before the launch happens. Last but not least, for new or problematic projects, be sure to conduct a lessons learned session with your customers some time within the first two months after launch.

In terms of project personnel, the framework identifies the role of project director and project manager. The project director is the IT party responsible for project delivery and the overall coordination of internal and external IT resources. He or she works hand-in-hand with the customer to ensure that project deliverables are in keeping with the customer's requirements. The project director has final say on decisions within the project team. Preferably, this assignment will fall to the party who will own the IT application or service once it is in production, and therefore someone with a vested interest in the project's success. The IT project manager is staff to the project director. This support person develops and maintains project commitment documents and plans, facilitates and coordinates project activities, carries out business process analysis, prepares project status reports, manages project meetings, records and issues meeting minutes, and performs many other tasks as required to ensure successful project delivery. Many IT organizations merge these two roles, but it is a good idea to separate them and indeed, to establish a centralized project management function (i.e., the PMO) with its own the support staff (i.e., project managers and business analysts).

The framework employs partner provider as shorthand for any technology specialists — internal or external to the IT organization — who contribute to the project's delivery but who are not necessarily involved in day-to-day project operations. For example, network services personnel may serve as partner providers to a development team delivering a new Web site. Working clients are customers assigned to the project by the executive sponsor (i.e., the party funding the project). Working clients bring business process expertise to the process and also serve as operational surrogates for the executive sponsor. Although the participation of working clients is often essential to a project's success, avoid awarding even these customers project manager or project director status because they often lack the necessary technical knowledge and experience in managing IT personnel and vendors.

As the preceding comments suggest, clearly defined roles and responsibilities are critical to the success of any project. However, there is one remaining role: process guardian. Actually, this is not a formal project management role, but it is important all the same. In any IT organization faced with the day-to-day management of multiple projects, there must be a person or office dedicated to overall compliance with best practices. This function would include the design and maintenance of project management forms, reports, and tools. The tracking of project management resources and interproject dependencies (in terms of people, expertise, the time of product and service releases, and so forth) would also fall to this role. Similarly, someone must own the master scheduling and monthly operational reporting activities discussed later in this chapter. Lastly, some party should follow up with project teams before, during, and after their initiation to ensure that they are following the standards set by IT in dealing with customers, measuring performance, and delivering results.

Obviously, I am referring here to the PMO and its manager. No one person can take on the responsibility for all of these activities. Rather, the manager of the PMO will work with his or her team to establish a consensus around best practice. No doubt executive IT management will contribute to and bless these standards. Thereafter, it becomes the PMO's role to embed these practices into the ways that the IT organization conducts its business. Although in some organizational designs, the PMO may staff the project management and business analyst functions within project teams, this may not always be the case, even when IT has its own PMO. However, regardless of IT's staffing design, the PMO's most profound impact is in the area of process management, where the team works to influence behavior across the greater IT organization. Through such practices, the methods and frameworks advocated in this book can work their way into the culture of the IT team and, over time, make a difference in overall team performance.

Lastly, the project management framework identifies process phase deliverables. Some of these, such as SLAs, have been addressed elsewhere in the text. Others, like project plans and project budgets, are self-explanatory and therefore require no further elaboration. Those few remaining project artifacts that may be unfamiliar include the project commitment document, project scorecards, and monthly operations reports. These particular project management components require special handling because of their unique but highly useful qualities and because they get little or no notice in other authors' treatment of the process. As the reader will learn, these particular elements can add to the quality of the IT project management experience and its successful outcomes.

THE COMMITMENT PROCESS

Initial project scoping is key to the subsequent steps in any project management process. All too often, projects are pursued without a clear understanding of the associated risks and resource commitments, leaving both the working clients and the project team exposed to the uncontemplated twists and turns of project execution and without a constructive understanding of their respective roles and responsibilities. Similarly, sometimes the most important information in framing an IT project — including the team's operating assumptions, the project's dependencies, and the customer's measures of success — is left undocumented and remains a topic of only brief conversation. If the various handoffs among team players also go unarticulated, these matters leave process participants without a clear sense of the process workflow and their respective responsibilities. Any of these traits can spell disaster for the project team and its project delivery efforts.

To avoid these and other contributors to delivery problems, IT project teams should embrace a commitment process that ensures a well informed basis for action. A framework for commitment management follows.* Like the other illustrations in this volume, this methodology's application should be balanced against the needs of the occasion. For example, if the project

---

* This section of Chapter 5 focuses on the commitment management process. The document for that process is templated in Appendix H and is also available in an electronic version. See *The Hands-On Project Office*, http://www.crc-press.com/e_products/downloads/download.asp?cat_no=AU1991, chpt5~5~commitment~template. For an example of the commitment document in use, chpt5~6~commitment~example.

in question covers well trodden ground, less rigor is required than if the envisioned project blazes trails into hitherto unexplored territories. The commitment management process itself forces the project team to ask the very questions that will help it to determine the best course of action. This framework and checklist also ensure that the team addresses all the elements of risk and uncertainty surrounding the envisioned project.

From the outset, no project should proceed without an executive (business) sponsor, at least one working client, and an identifiable source of funding. The executive sponsor's role is to ensure the financial and political support to see the project through to completion. He or she owns the result and is therefore the project's most senior advocate. If the project in question happens to be sponsored by the IT organization itself, the chief IT executive will serve as its sponsor, and the IT manager who will own the system or service once it is in production will act as the working client. As mentioned earlier, working clients operate as the liaisons and day-to-day project team members from the line-of-business side of the house. They bring intimate knowledge of the business requirements driving the project, as well as a detailed understanding of the role that IT enablement will play within their business units. Lastly, the sponsor and the working clients must bring to the table the resources required for project delivery, including the funding for hardware and software purchases, hiring external consultants and contracts, and if appropriate, paying the IT organization for its participation. Note that I take it as a given that the planning process has approved the project in question for consideration by IT.

With these prerequisites in place, the project leadership — typically the project director, the project manager, the working clients, and technical subject-matter experts — will define the scope of the project. The first thing for the team to ask itself is, "Do we know enough about the project at hand to define its parameters confidently, including the risks involved, time and cost requirements, the skills required, and the technologies to be employed?" Remember that your team must commit to the project at some point. It should not do so unless it has a clear sense of what is involved operationally and technically to meet the customer's requirements. If, for example, the project is one with which the team has had experience, team members may come to the table with some confidence. But if they are treading on new ground, my recommendation is that they "chunk" the project, starting with an analysis phase. The outcome of such a first step would be to establish the knowledge to create a solid set of customer requirements, a thorough risk analysis, a project plan, and a budget for the remaining phases of the project. With this information in hand, the team is now positioned to commit to the actual delivery of an IT solution. By drawing on past experiences and by exercising due diligence up front, the project team will find itself in a much stronger position from the outset to manage customer expectations and to mitigate project risks.

Many IT projects are routine, but some are transformational. Anyone who has had the dubious honor of leading the implementation of an enterprise resource planning (ERP) system, such as PeopleSoft, Oracle, or SAP, knows exactly what I am talking about. Here, the scope of the project touches most key aspects of the business and requires major process change across the enterprise to have the desired impact on corporate performance. In this context, project leadership takes on a whole new meaning. Before entering into such an assignment, business and IT executive managers must ask themselves, "Have we prepared the organization for the degree of change driven by this commitment to information technology? Do we have the right people in place to lead the change process?" and so forth. From these considerations will emerge any number of more granular issues, as illustrated in part by Exhibit 6.*

As you can see from this excerpt, preparing the enterprise for a truly transformational IT project is no trivial matter and requires the participation of the most senior layers of management to succeed. Although there may be a legion of other challenges facing the team, the absence of effective leadership in terms of organization change and business process reengineering can doom a large-scale IT project from the outset. As a next step in framing the team's commitment, the project leadership should define the business problem or opportunity driving the proposed investment of IT resources. This may seem a trivial activity, but it is surprising how disparate the initial conversation on this subject may become. It is essential that the team start from a common base of understanding about the project's rationale and purpose. To that same end, the author likes to walk project teams through the value template shown in Exhibit 7 so that everyone involved can appreciate the benefits of a positive delivery outcome.

Avoid exaggeration. Remember that your chief financial officer (CFO) may view this document, and if you claim that the project's delivery will result in a major increase in revenue or a cost savings, the CFO will hold you and your line-of-business partners accountable for that positive impact to the bottom line. Few major IT projects save money, because they invariably incur added costs for the upkeep of the new systems that offset

---

* The entire questionnaire is templated in Appendix F and is also available as an electronic form. See *The Hands-On Project Office*, http://www.crc-press.com/e_products/downloads/download.asp?cat_no=AU1991, chpt5~7~project leadership readiness~template.

Exhibit 6   Project Leadership Questions — An Excerpt

THE LEADERSHIP COMPONENT:

- Who is leading the project? ____
- Is that person the most appropriate responsible party to ensure success? ____
- Does the leader genuinely believe in the project and want it to succeed? ____
- Does the leader have the necessary skills for success?
  Demonstrates genuine respect for people throughout organization ____
  Listens well ____
  Demonstrates political savvy ____
  Possesses connections with others critical to the project's success ____
- Has project leadership been legitimized, as appropriate? ____

ORGANIZATIONAL LINKAGES:

- Has a compelling case for change been made? ____
- Has the case for change taken into consideration other changes already under way around the organization? ____
- Has the project team completed a stakeholder analysis, measuring the relative impact on stakeholders? ____
- Is there a clear and convincing linkage between the proposed project and the enterprise's strategy, goals, and objectives? ____
- Does this discussion include a consideration of the needs and concerns of long-standing members of the enterprise who may view proposed changes as a critique of past performance and contributions? ____

EXPECTATIONS:

- Has the project team diagnosed employee attitudes toward the proposed change? ____
- Will they participate in the change process? ____
- Has the team determined the optimal time for project kickoff? ____
- Has a clear message gone out to the enterprise concerning realistic project accomplishments? ____
- Has sufficient information about potential payoffs been disseminated so that employee expectations are high enough to ensure general involvement? ____

REWARDS:

- Have arrangements been made to reward participants for their
  Time? ____
  Ideas? ____
  Commitment? ____
- Has the project's impact on existing reward systems been analyzed? ____
- Has the reward of participating
  Been recognized? ____
  Been taken into account for its impact on nonparticipants? ____

Exhibit 7   Project Benefits Matrix

Business Improvement	Major	Minor	None	Business Value Statement (in Support of the Improvement)
Increase revenue	_____	_____	_____	_____
Decrease cost	_____	_____	_____	_____
Avoid cost	_____	_____	_____	_____
Increase productivity	_____	_____	_____	_____
Improve time-to-market	_____	_____	_____	_____
Improve customer service/value	_____	_____	_____	_____
Provide competitive advantage	_____	_____	_____	_____
Reduce risk	_____	_____	_____	_____
Improve quality	_____	_____	_____	_____
Other (describe)	_____	_____	_____	_____

savings elsewhere. Furthermore, although IT may bear the cost of the new service, cost savings may come from the customer's side through business process changes over which IT has no influence. On the other hand, cost avoidance, risk mitigation, improved customer service, and even competitive advantage may all be positive attributes associated with your project. Think through the case for each benefit and justify it in the space provided on the form.

With a common view of the overall project vision and value in place, the time has come to detail project deliverables, including those that are essential for customer acceptance, those that are highly desirable if time and resources allow, and those that are optional (i.e., the project may be acceptably delivered without these components). Next, indicate elements that are excluded from the scope of this project (but that may appear in future, separately funded phases of the project). This is a particularly important step. Do not hesitate to state the obvious. You will probably be amazed at what your customer assumes is included within your project's scope. By clearly stating what is in and what is not, you will avoid unpleasant surprises as your delivery date approaches. Some project deliverables may not be apparent to the team. For example, if your task is to replace the technology in a laboratory facility, you may not consider facilities alterations part of your mandate. Similarly, in delivering a software application to a business unit, you might easily overlook the desktop or information security implications of that change. To avoid such oversights, ask your PMO to develop checklists that your project teams may then regularly employ as part of their scoping exercises.[*]

Given the project's now agreed-upon deliverables, the team should assign critical success factors for customer satisfaction based upon the following vectors of measurement: scope, time, quality, and cost. These metrics must be defined in terms of the particular project. For example, if a project must be completed by a certain date (e.g., recent efforts to make systems year 2000–compliant), time rises to the top of the list. If time grows short, the enterprise will either adjust scope, sacrifice quality, or add to costs to meet the desired date. Similarly, if the scope of a project is paramount, its delivery date might be moved out to allow the team to complete the commitment. If time is also an important factor in delivery, more people could be added to the project to deliver it in scope and on time.

In a recent instance, the author was asked to develop an enterprise data mart for a business partner. The data elements, date of delivery, and resources allocated for the project were all fixed. However, the sponsor was willing to give on the quality of the data in the data store. He agreed that data cleanup would follow project delivery. As with many other aspects of the commitment process framework, the importance here is to ensure that a thoughtful discussion ensues and that issues are dealt with proactively rather than in a time of crisis. Obviously, the discussion of critical success factors must take place with the working client(s), creating a golden opportunity to set and manage customer expectations.

Because no major change to an IT environment is without implications, the commitment process must also identify any major impacts to other systems and services that will result from the implementation of the envisioned IT solution. For example, if a new application requires network infrastructure or desktop platform upgrades, these must be noted in the commitment document and their implications carried over more tangibly into the project plan and budget. Similarly, if a new information system requires the recoding of older systems or data extracts from enterprise systems of record, these impacts must be documented and factored into the project plan. As part of the analysis phase of your project, you must run these implications to ground and factor them into your plans. The

---

[*] See, for example, *The Hands-On Project Office*, http://www.crc-press.com/e_products/downloads/download.asp?cat_no=AU1991, chpt5~9~facilities project delivery~template, chpt5~10~Infrastructure Questionnaire for Projects~template, and chpt5~11~Security Questionnaire for Projects~template. In a similar vein, the following tool, originally developed by Pat Laughran, is a useful reminder of the pre-launch and launch steps for an information system or service; see *The Hands-On Project Office*, http://www.crcpress.com/e_products/downloads/download.asp?cat_no=AU1991, chpt5~14~system launch checklist~template.

sooner you catch them, the less costly they will be to your project. This area is one where the PMO team can be of considerable assistance. PMO staff are more likely to spot linkages among projects and the ripple effects of system and platform changes. Encourage them to employ their informal networks to ferret out the facts.

What often gets a project team in trouble is not what is documented but what goes unsaid. For this reason, the commitment process should require the team to explore project assumptions, constraints, and open issues. Because most IT professionals tend to be internal thinkers, this can be a formidable task. It therefore falls to the project's director or manager to draw out from the team and make explicit the inferred operating principles of the project, including the roles and responsibilities of project participants (especially internal and external IT partner providers), how project activities should operate, what tools and technologies are to be employed, and how key business and technical decisions governing project outcomes will be made. All projects operate under the constraints of time-resource availability and dependency on the actions of others. For example, a particular project may be constrained by the unavailability of a particular technical specialist or by delays in the arrival of computer hardware and software. These factors may directly impact project outcomes but are out of the team's direct control. Make them explicit, so that the customer appreciates the risks to the project should they occur. Once they are made explicit, the project manager should track these issues to determine if and when they may adversely affect project outcomes. PMO staff can then work together to mitigate the associated risk, or at least to make it more visible to IT senior management and the affected project teams.

Open issues are different from constraints in that these elements can and will be addressed eventually by the project team. They appear in the commitment document to remind all those involved that if they remain open, these issues will adversely impact delivery. Thus, the open issues section of the commitment document is a parking lot for major follow-up items. Here, too, have the project manager oversee the list and ensure that all open items are closed in due course over the life of the project. The project manager or the project director should share the status of these items with the customer on a regular basis as part of initial expectation setting and subsequent project reporting.

The two remaining components of the commitment process are those elements that reflect project risks and those elements that itemize the project's specific resource commitments. Exhibit 8 shows an illustrative risk management matrix.

In completing a commitment process, the project team should identify the major risks in pursuing the assignment. The table identifies risk categories and provides room for a more detailed description of a particular risk and

Exhibit 8 A Risk Management Matrix

Potential Risk	Description of Risk	Resolution
Technology		
Financial		
Security		
Data integrity		
Business continuity		
Regulatory		
Business requirements		
Operational readiness		
Other (explain)		

its mitigation. For example, a project technology risk might entail introducing a new or untried technology into the enterprise's IT environment. A way to mitigate that risk would be to involve the vendor or some other experienced external partner provider in the initial installation and support of the technology. If the envisioned project solution requires clean data to succeed, the project plan could include a data cleanup process. If business requirements are not documented, the project could call for business analysis and process engineering work to get at those requirements. All in all, the team must be honest with itself and its customer in defining and dealing with project risks. Keeping risk statuses in the commitment process ensures that they remain visible and that those responsible are held accountable for their mitigation. Here again, the PMO's project manager will work in partnership with the project director to limit the undertaking's risk exposures.

The primary categories of IT project risk may be defined as follows:[*]

- Technology risk — the project calls for a new product or an untried technology supplier, a product with which the enterprise's IT organization has no expertise, or the untried integration of that product with other products already in place within the enterprise's IT environment
- Financial — the project is not sufficiently funded either for the initial deployment or in terms of ongoing operation and support of the implementation

---

[*] A more comprehensive risk management tool can be found in Appendix G and on the accompanying Web site. This tool allows the reader to rate and score the various risks associated with a particular IT project. When one is faced with options, the tool allows the reader to compare choices from the standpoint of risk management. See *The Hands-On Project Office,* http://www.crcpress.com/e_products/downloads/download.asp?cat_no=AU1991, chpt5~8~risk management~template.

- Security — the project possesses information security risks or exposes the enterprise's information assets to unwarranted access
- Data integrity — the existing quality of enterprise data could jeopardize the project, or the project's implementation could adversely impact the data integrity of other enterprise information systems
- Business continuity — failure to deliver the project will adversely impact the enterprise's ability to conduct business
- Regulatory — failure to complete the project jeopardizes the enterprise's regulatory situation, or delivery of the project may compromise the enterprise's regulatory situation
- Business requirements — the customer has not provided clear and complete project requirements
- Operational readiness — either the business unit or the IT organization is not prepared to operate the envisioned IT solution (e.g., there are no business processes in place; the staff needs training)

The project team, and in particular the project manager, must be attuned to any of these risks as they apply to the assignment, tracking and addressing these issues as need be.*

The document must indicate project resource needs in terms of people, time, and funding. Here, the tool allows for an optimistic set of best-in-class estimates; formal, planned estimates (i.e., expected outcomes as documented in the project plan); and contingency allowances. Although some projects come in ahead of schedule, more often than not, they come in late and at a higher cost than first estimated. Give your team some wiggle room by building in a realistic contingency. Bear in mind that although it is not a good practice to pad budgets and timelines, it is important to alert your customer to the reasonable dimensions of cost and time overruns that the project may experience.

From the standpoint of people, the document must explicitly name names and define roles and responsibilities (including the skills required). Exhibit 9 shows an illustrative list of project roles. The project director must ensure that a real person or people are assigned to each project role and that chosen IT staffers understand and agree to their assignments. See Exhibit 9. These commitments cannot occur without a delineation of the other two resource elements:

---

* The accompanying tracking tool will help organize project issues and risks as they arise. See *The Hands-On Project Office*, http://www.crcpress.com/e_products/downloads/download.asp?cat_no=AU1991, chpt5~12~Project Issues and Action Items~template, and chpt5~13~Project Issues and Action Items~example.

Exhibit 9 Roles and Responsibilities Matrix

Role	Name of Associate	Responsibility
The Core Project Team:		
Executive Sponsor		
Working Client(s)		
Project Director		
Project Manager		
Business Analyst		
Application Lead		
Systems Lead		
Data Management Lead		
Infrastructure Lead		
Customer Services Lead		
Internal and External Partners:		
Vendor-based Project Management Support		
Technical Architect(s)		
Business Process Architect(s)		
Creative Development/UI		
Development		
Training/Documentation		
QA/Testing		
Infrastructure		
Security		
Other Partner Provider(s) – Hardware/Software:		

- Skills, time, and duration of the commitment for each internal staff person
- Associated funding for hardware, software, contract labor, consulting, and so forth

These details will come from the project plan that accompanies and complements the commitment document. In the plan, activities are appropriately detailed, along with the duration and performer of each task. The plan tells each partner provider what is required of his or her team. It is the responsibility of these managers to ensure that they do not overcommit their own personnel. If the IT organization operates some sort of resource management database or tracking system, this may be easily accomplished. Otherwise, it rests with individual managers, as aided and abetted by the PMO, to keep things straight.

The commitment document is a key tool in project management process. Through its disciplined use, the team captures all the aspects of risk and ambiguity surrounding the envisioned project. It affords approaches to resource management and risk mitigation. Lastly, through the signoff process, the commitment document acts as a formal contract binding the players to their roles and responsibilities within the delivery process. During the course of delivery, things will change. Nevertheless, the commitment document serves as a point of reference, ensuring that all the key questions are raised at the initiation of a project and that the team, supported by the PMO, addresses all outstanding issues in due course. A number of other useful tools, most cited in the footnotes and housed in this book's accompanying Web site (http://www.crcpress.com/e_products/downloads/download.asp?cat_no=AU1991), can assist the project team in reaching a successful conclusion.

When viewed in its entirety, the commitment process leaves nothing to the imagination of the project team and those they serve. The commitment document makes explicit what is to be done, why the project merits resources, who is responsible for what, and what barriers lie in the path to success. The project plan details how team members will execute their respective assignments. Together, these documents form a contract that aligns resources and provides for a common understanding of next steps, roles, and responsibilities. The metrics for successful project delivery are few and simple: Did the project come in on time and within budget? Did it meet customer expectations? To answer these questions, simply run actual project results against the project's commitment document and plan.

This framework makes for a good beginning, but it does not ensure the success of the project. Being true to the commitment process addresses proactively many of the problems that might otherwise befall a project. Still, miscommunication and invalid assumptions can derail even the most effectively started projects. The concluding section of this chapter closes this gap by offering an approach to project performance measurement and reporting. Here, the PMO and its agents in the field — its project managers — hold the keys to success.

PROJECT DELIVERY — MEASUREMENT AND REPORTING

Considering all of the work that a typical IT organization is expected to deliver each year, it is easy to see how even major commitments may be overlooked in the furious effort to get things done. To avoid this pitfall, it is incumbent upon IT to clarify its commitments to all concerned. Next,

IT management must ensure compliance with its commitments. The ongoing project management process forces the team to relate actual accomplishments to plan. If the IT organization employs customer relationship executives (CREs) as customer liaisons, these folks will regularly visit their assigned customers to review their SLAs and any outstanding project delivery commitments with the IT organization. The goal of these discussions is to assess overall customer satisfaction with IT performance. These talks, however, will rarely dwell on the factual details without an effective and efficient mechanism for conveying that information. To this end, the author employs operations report data, as detailed in Chapter 4, and a collection of one-page project scorecards that, in single snapshots, convey all the essential information about particular project statuses.*

The scorecard is a monthly view of project work that includes a brief description of the project, the customer value proposition, a list of customer and project team participants, this month's accomplishments and issues, a schematic project plan, and a Gantt chart of current project phase's activities. Each scorecard is color-coded as follows:

- Red — the project is in trouble but its problems are beyond the control of the project team (e.g., a vendor is late with product delivery, or another project, upon which the current project depends, is seriously behind schedule).
- Yellow — the project is facing difficulties, but the team has those problems under control.
- Green — the project is on time and on budget and does not face any serious difficulties.
- White — the project was completed during the past month and was signed off on by the sponsor and the working clients.
- Purple — the project is on hold or otherwise pending.

See Exhibit 10.

Because it is designed to be customer-centric, the scorecard's features make it accessible to all. The objective and value statements align the project work with the goals and objectives of the enterprise. The monthly highlights report on current accomplishments relevant to customer delivery, just as the issues statements speak to the immediate barriers to success. The summary project plan tells the customer what he or she really wants to know: "What are the key project milestones, and when will I see

---

* The project scorecard template may be found in an electronic form at *The Hands-On Project Office*, http://www.crcpress.com/e_products/downloads/download.asp?cat_no=AU1991, chpt5~15~project~scorecard~template. For examples of the scorecard as applied to real-life projects, see chpt5~16~scorecard~examples.

Project Delivery and the Project Management Life Cycle • 141

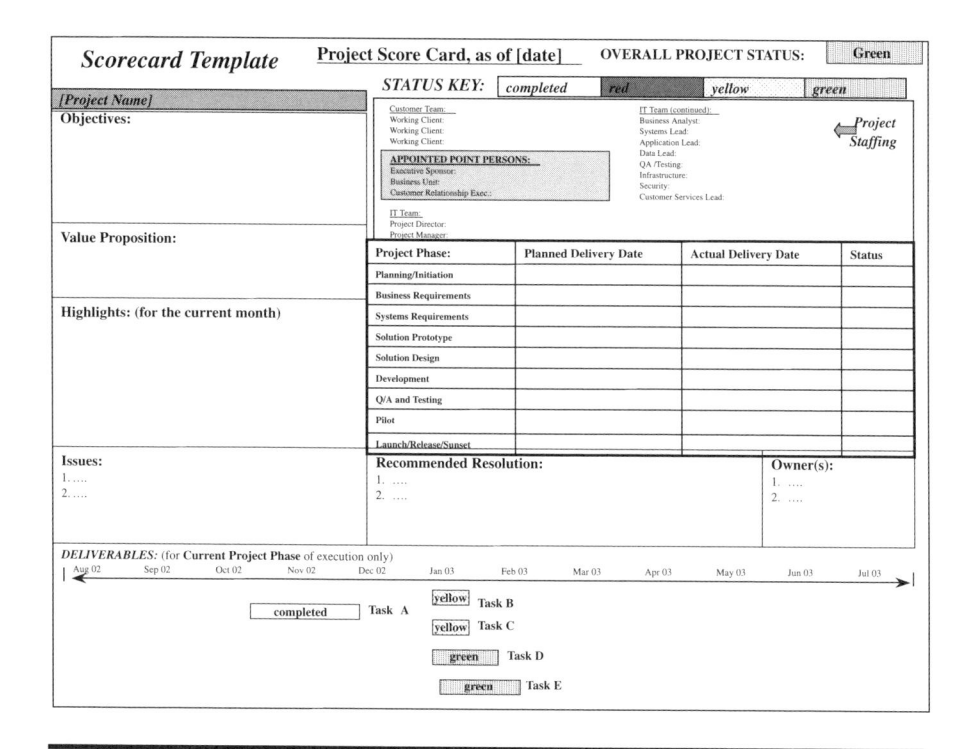

Exhibit 10 Scorecard Template

results?" Lastly, the Gantt chart at the bottom of the scorecard details the timeline for the current phase of activity. In short, the scorecard covers all the information that a sponsor, working client, or CRE needs and wants to know about the status of a project in that business unit's portfolio. If more is required, the CRE can work with the project team to produce the added information in real time.

Typically, the project manager prepares the scorecard with input from the project director. This is an important distinction. Bear in mind that although the project manager supports the project director, he or she remains independent. The expectation is that the project manager will bring his or her objectivity to bear when preparing the scorecard and will be more aware than the project director of other projects in the overall IT portfolio that may be adversely impacting its delivery. If the director and manager disagree on the status of their project, the director has the last word. In practice, such conflicts are more often the exception than the rule. More to the point, the project manager and his or her PMO colleagues exercise considerable influence over these matters, helping the project director to view and present team accomplishments honestly and to assist in addressing issues as they arise.

Customers are not the only consumers of scorecard contents. The IT management team has an interest in the status of all IT organization project work on three different levels. First, there is the immediate concern about the status of IT's commitments to its customers. Because the management team also serves as the CRE cadre for the IT organization, it needs to know about the general health of all projects under way, completed over the past month, or in the pipeline. When they present updates to their customers, CREs should appear well informed about the life cycles of each of these assignments. For most CREs, this is a lot of information to digest. Fortunately, their project scorecards provide the right level of knowledge for such exchanges. These documents will also highlight problems that may be researched for a more detailed understanding before any meetings with sponsors and working clients.

The second level of interest in scorecard data is what these updates may suggest concerning the IT team's overall service commitments. As the scorecards report on the sunsetting of superannuated systems, the introduction of new products and services, or the integration of new classes of end users, all of these events have implications for the infrastructure, support, and training teams. Furthermore, because as a practical matter it is nearly impossible for IT line managers to engage simultaneously in all ongoing IT projects, scorecards serve as an early warning system for those IT partner providers. Upstream providers will learn how delays in their own assignments may impact others. Downstream consumers will find out about developments that may in turn influence their ability to meet their own project deadlines and may encourage them to plan around these delays or pressure those in their critical path. Here, too, the project managers take responsibility for these interproject team discussions and negotiations, in effect serving as an early warning system in this web of commitments.

Finally, as a package the monthly scorecards act as a bellwether for how the IT organization is handling its project work. Because there are so many projects to consider at any one time, however, IT management requires a more aggregate view of the unit's project delivery performance. To this end, the author has relied on a single, integrated reporting process, called the monthly operations report (see Chapter 4), to capture key IT accomplishments and performance metrics and to serve as a wrapper for the packaging of the monthly scorecards.*

---

* The projects wrapper of the monthly operations report is pretty modest compared to that of the service level delivery side of the document. What gives the project side its depth is the master project schedule and the associated scorecards. See *The Hands-On Project Office,* http://www.crcpress.com/e_products/downloads/download.asp?cat_no=AU1991, chpt5~17~monthly project status report~template and chpt5~18~monthly project status report~example.

As its name implies, the monthly operations report is a regularly scheduled activity. It is designed to serve the needs of IT management, keeping customer delivery at the forefront of everyone's attention and holding IT personnel accountable for their commitments. The report captures qualitative information from each IT service delivery unit (e.g., help desk, training center, network operations, production services, and so forth) concerning issues and accomplishments for the prior month. It then brings together all of the IT organization's most current project updates. Unlike service delivery performance, project delivery may be somewhat complicated to capture on a monthly basis because projects do not necessarily lend themselves to quantitative measures or to regular customer satisfaction surveying. Nevertheless, the report contains two sets of documents that IT management will find useful.

The first is a complete set of project scorecards for the month. I have already dealt with the benefits of these documents individually. Collectively, they afford a portfolio view by the business unit being served or by the IT organizational unit or project team providing service. From the management point of view, these documents prove invaluable when managing customer relationships, proactively addressing customer problems and complaints, and planning follow-on IT investments with individual business units or for the greater enterprise IT infrastructure. The scorecards also alert management to internal IT team problems — both technical and interpersonal. Managers may choose to seize the opportunity afforded by a particular scorecard to recognize and reward or mentor project team members. Reported results may also prompt IT management to make changes in resource allocations, staffing assignments, internal processes, or the services of the PMO.

The second major document type included for project management in the monthly operations report is the project master schedule, as prepared by members of the PMO. The project master schedule groups projects by customer portfolio and then by interproject dependencies.[*] The schedule shows the duration of each project and its status (white for completed, green for on schedule, yellow for in trouble but under control, red for in trouble, and purple for pending). The schedule also reports the names of the project's director and manager, as well as any major issues or dependencies impacting project delivery. In larger organizations, the schedule may even serve as a contact directory for those seeking information about a particular project. Thus, within a few pages,

---

[*] See *The Hands-On Project Office*, http://www.crcpress.com/e_products/downloads/download.asp?cat_no=AU1991, chpt5~19~master project schedule~template and chpt5~20~master project schedule~example. For a hardcopy version of the tool see Appendix I.

IT leadership can see all of the project work under way at any given time, including projects that are in trouble, critical path bottlenecks, and the names of staff members who might be overcommitted. The presentation is simple and visual. The report's scorecards align with the project master schedule to provide an economical yet comprehensive picture of all project activity within the IT organization.

Although the monthly operations report process primarily serves the needs of IT management, appropriate sections of the report, as well as individual project scorecards, should be shared by the unit's CREs with their respective customers. Through routine meetings with customers, the CRE or the project director and manager keep their sponsors and working clients informed about project statuses. As they inform the customer about project issues and accomplishments, these IT organization representatives manage customer expectations. In brief, the process keeps all projects visible and everyone on their toes. Because project statuses are also reviewed at operations review meetings, no one wants to come forward with bad news, but all must be encouraged to speak up honestly. Here, the independence and objectivity of the PMO may be exercised to good effect by asking PMO personnel to draw out the lessons learned from troubled projects. Bear in mind that the focus of this process is continuous improvement and the pursuit of excellence in customer delivery. Blame is never assessed individually because the entire IT team is held accountable for the success of the whole. These are all powerful outcomes to be gained by a modest and well distributed investment of time in self-evaluation and timely reporting.

THE ROLE OF THE PROJECT MANAGEMENT OFFICE IN PROJECT MANAGEMENT SERVICES

Although all of the processes described in this chapter have some merit and may even work for your own organization, there is no simple way to implement such a vast array of practices. Admittedly, the methods outlined here carry with them a certain level of administrative overhead. On the other hand, the cost of developing and maintaining these processes will prove insignificant when balanced against the payback — high-quality customer relationships and repeatable success in IT project delivery. But the question remains: "How does an IT organization achieve the level of self-management prescribed in this chapter?" Here is my answer: "Establish a project management office to ensure the synchronization of service level and project delivery."

Because many of the tasks cited in this chapter must be accomplished in the course of project delivery anyway, it is more efficient to establish a center of excellence in these skills that in turn ensures broad-based IT

Project Delivery and the Project Management Life Cycle ■ 145

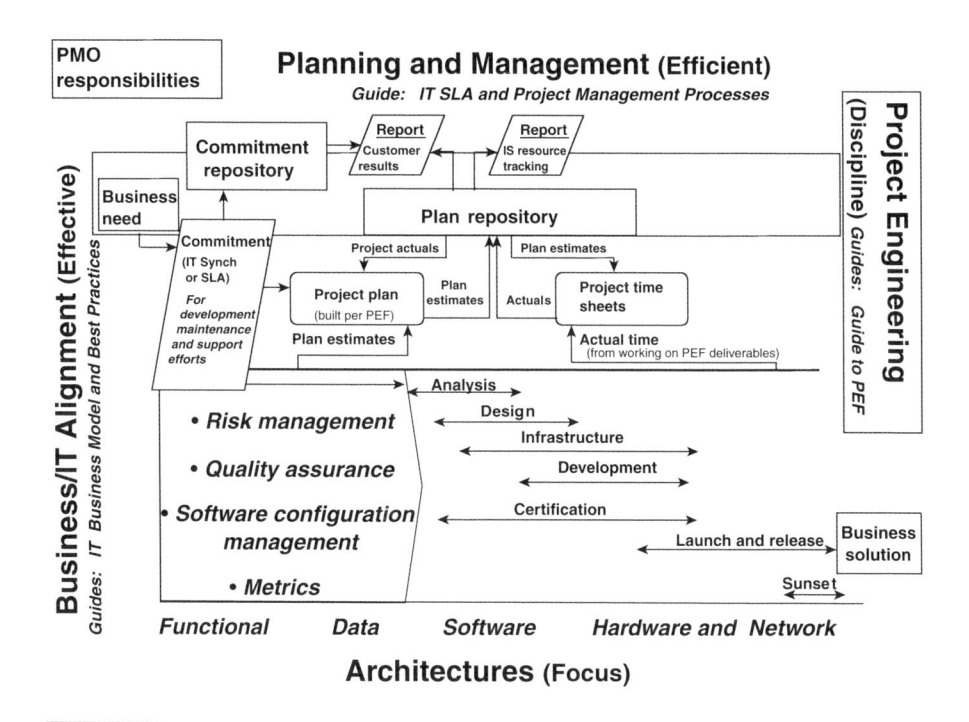

Exhibit 11   A Project Delivery Framework for the PMO

compliance with best practices. Furthermore, nothing about the processes described herein is static. As IT and its customers learn from the use of these processes, they will fine tune and adapt them based on practical experience. PMO personnel will become the keepers and the chroniclers of this institutional learning and its knowledge management by-products. Because they operate independently of any particular IT service delivery or project team, PMO staff are in a position to advocate for and monitor the success of commitment management and service/project delivery processes.

In closing, Exhibit 11 offers one last image for framing the role of the PMO in the context of overall IT project delivery. Although this illustration is rather busy, it is also accurate in its conveyance of the various dimensions of the PMO's contribution to the health of the project management process.*

Along the left axis are references to business and IT alignment. As you may recall, Chapter 3 identifies a number of supporting roles for the PMO in the fashioning of the IT organization's annual plan. These

---

* For an electronic version of the framework, see *The Hands-On Project Office*, http://www.crcpress.com/e_products/downloads/download.asp?cat_no=AU1991, chpt5~21~project engineering framework~model.

activities provide PMO personnel with a comprehensive and timely understanding of IT team goals, objectives, and priorities for the coming year. In short, the process gives them insight into the right things to do. Proceeding to the top axis and IT's planning and management processes, PMO personnel assist in the crafting of SLAs and commitment documents that in turn define what, how, and by whom services and projects will be delivered to IT's customers. Here, the PMO serves as the keeper of past IT team experience, the promoter of best industry practices, and as the scribe and coordinator of IT delivery and customer relationship management processes.

Along the bottom axis is the project delivery life cycle, when PMO project managers and business analysts participate directly in delivery. Here, they are not only members of the support team but also directly creating customer value through their efforts. Furthermore, the PMO will continue to monitor overall life-cycle management and recommend process improvements as these emerge from the project delivery teams. Lastly, the bottom axis suggests how the PMO — as the promoter of standard practices; of the reuse of information, tools, templates, and technical components; and of adherence to IT's technical architecture — helps to keep the rest of the IT organization focused on the right things and to avoid reinventing the wheel with each new assignment. All in all, Exhibit 11 nicely summarizes what the PMO can and should do for the IT organization it serves.

All in all, the frameworks and tools outlined in this chapter should prove useful in controlling initial project scope and preventing project scope "creep." Furthermore, my communication and reporting approaches will ensure that all project stakeholders keep abreast of development and have opportunities to contribute to project discussions.

6

COLLECTING AND CAPTURING BUSINESS REQUIREMENTS FOR IT PROJECTS

INTRODUCTION

At several points, this volume has alluded to business analysis as a role within the project management office (PMO). The business analyst is integral to project delivery and continuous process improvement — both among the customers served by the IT organization and within IT itself. In fact, changes within the enterprise and IT go hand-in-hand. Almost invariably, when new technologies are added to established or emerging business practices, those processes must change to reap the full benefit of IT enablement. Similarly, the IT organization must adapt itself to the additional products and services in its portfolio of deliverables. Within this grand scheme, the business analyst's role supports the design, development, and implementation of change in a number of ways:

- Documenting existing business processes and customer uses of IT
- Developing and documenting process flows in terms of a particular technology-enabled solutions (i.e., functional specifications)
- Developing project business and functional specifications and assisting in the drafting of technical specifications
- Working with the project manager to develop statements of work, project schedules, deliverable descriptions, issue tracking documents, and management reports

- Preparing test scripts and quality assurance scenarios for the testing and release management processes
- Facilitating and supporting business process reengineering within customer departments and perhaps within the IT organization itself

Key to this work is gathering knowledge about how the enterprise serves its customers. As an outcome of this process, business/IT project teams can establish the particulars for reengineering and IT-enablement undertakings. Look around your organization and you will see that many of your colleagues are regularly engaged in the collection and analysis of customer business requirements. For instance, service delivery personnel must gather customer requirements for their SLAs and performance measures. Project management teams get deeply involved in the documentation of business requirements as part of the analysis and design phases of any project. Thus, the process of business requirements gathering is important to most of us in IT. On the one hand, the quality of this work determines how well you understand your customer's need for and commitment to process change; on the other hand, the effort establishes the metrics for customer satisfaction in the delivery of the IT systems that will help to enable those process changes. By adhering to a set of thorough and accurate customer requirements, IT personnel will more properly align product and service delivery with their customers' needs and expectations.

There are many sound approaches to the process of business requirements gathering. Some of these are cited in the selected readings section in Appendix K. The author's own view of requirements gathering reflects his interests in business process reengineering and knowledge management. I blend these two disciplines in my approach to the task. To frame this discussion, let us consider the context for the work of business analysis.

Each enterprise's business processes dictate their own information needs and associated IT requirements. The typical enterprise has no more than six or so key business processes. Within these large operating realms, however, there may be any number of information management subsystems in play, each with its own demands for IT. At any given time and as part the allocation of IT resources and effort, the enterprise will focus on some subset of these processes, looking to enhance their capabilities or even to completely reengineer how they impact the business unit and its customers. A sample set of key business processes within an enterprise might include the following:

- Lead generation — the processes of market and competitive analysis, market awareness and branding, customer prospecting, and prospect profiling and identification
- Sales cycle management — the process of converting prospects into customers

Collecting and Capturing Business Requirements for IT Projects ■ 149

- Delivery management — the manufacturing process,* including customer and engagement management
- Distribution management — the process of product or service delivery to the customer
- Financial management — the processes of accounting, accounts receivable administration, accounts payable administration, financial controls, and reporting
- Human resource management — the processes of payroll and benefits administration, staff recruiting and retention, and staff training and development

Note that these logical groupings of business processes capture most, if not all, process work within an enterprise. The nature of your particular business may require that you recast these representational groupings to better reflect your marketplace (e.g., a bank might refer to delivery management as service delivery, and a university might refer to the same grouping as program and classroom delivery), but the intent is the same. Within these headings, one may arrange all of the products and services that IT provides in support of enterprise business processes. As Chapter 2 and Chapter 3 have already discussed, external market forces and customer demand, as well as the internal drivers within the enterprise for process improvement, will encourage your business leaders to reevaluate their technology investments regularly. In some cases, they will recognize the need to enhance existing IT systems; in other instances, they will call for their wholesale replacement. When this occurs, line-of-business management will call upon the unique skills of the IT organization to help sort out the change process and to integrate or reintegrate technology with patterns of work.

As a starting point, business managers typically consult with their IT organization counterparts (or their IT CREs) to identify and select those IT projects of the greatest value to the enterprise. For example, if the enterprise has just expanded its sales force or faces problems with customer retention, the team might target the technology complex that supports sales cycle management for a face lift. Similarly, if the enterprise finds that it cannot compete in terms of the cost effectiveness of its manufacturing process, the team might choose logistics or distribution management for an IT-enabled makeover. Your organization should follow some rational process, like the

---

* In a mature service economy, such as that of the United States, the enterprise's deliverables to the customer are usually a set of services rather than tangible goods. Therefore, the reader should not limit his or her frame of reference to a particular subset of our service economy. The author's model applies to all types of for-profit or not-for-profit enterprises.

one offered in Chapter 3, to identify, select, and prioritize projects for the IT team. Eventually, project teams will be charted and the analysis will begin to determine how best to deploy information technologies in support of your business partners' requirements. Even during the commitment process detailed in the previous chapter, the project will need to conduct enough research to properly scope the assignment and to recruit colleagues with the appropriate skills for the work at hand.

In choosing an approach to requirements gathering and analysis, your team must take into account the culture of enterprise, the quality of the IT organization's current relationship with the business units in question, and the nature of the business process and set of technologies under review. No two situations call for the same approach. In most instances, however, your business analysts should gather the same sorts of information to establish replacement processes, including process workflows, business rules, the roles and responsibilities of process participants, existing points of IT enablement, and so forth. This chapter offers a systematic yet simple and straightforward way to decompose and document business process components for subsequent analysis. The benefits of this approach reside in its adaptable application to very different sorts of business activities and its ability to engage customers in the discovery process. If, as a result of this rapid-fire approach to requirements gathering, the team uncovers issues that call for more detailed research, analysts may devote more focused time to these components of the business process. In the end, you will find that perhaps 70 to 80 percent of your questions are answered through the thorough use of the author's template, complemented by other methodologies as needed.

PREPARING FOR BUSINESS REQUIREMENTS GATHERING

To begin, the project team should consider the nature of data to be collected, categorized, and managed. To this end, the author offers a simple, knowledge-components model that characterizes the categories of knowledge to be drawn upon at the outset of requirements gathering:

- Marketplace/customer knowledge — information about business barriers and opportunities, customer profiles, prospect profiles, market demographics, and so forth
- Content knowledge — information about your enterprise's actual products and services, performance history, staff competencies, intellectual and physical assets, and so forth
- Process knowledge — information about how your enterprise manages itself in terms of its key processes (e.g., solution selling, manufacturing, distribution), as mentioned previously

Much of this information will come from sources internal to the business process under study, but some must also come from upstream providers to that process and the downstream customers of that process. For example, as an IT team begins work on the design of a new manufacturing system, it may focus its attention on interviewing key participants in the manufacturing process. At the same time, the team must include input from the logistic people (upstream) who provide the raw materials for manufacturing and the warehousing and distribution people (downstream) who handle the finished product.

The actual sources of information will themselves vary from one business function to the next and will be greatly influenced by the culture of the business unit in question. For example, any business operation that is audited on a regular basis or must comply with a great deal of regulation, such as finance and accounting or human resources, is likely to have formal policies and procedures already in place. These documents would be an excellent starting point for project discovery work. As your business analysts begin their data-collecting activities, consider these factors in searching for documentary evidence about business processes:

- Information may originate from sources external or internal to the business unit under scrutiny by the project team
- Information may be explicit (recorded in some medium) or tacit (walking around in the heads of content or process experts)
- Information may concern the enterprise's marketplace and customers, for example:
  - Business barriers and opportunities
  - Customer profiles
  - Customer and market demographics
- Information may concern the content of the business, for example:
  - Product descriptions
  - Service descriptions
- Information may concern actual business processes, for example:
  - Solution-selling workflows
  - Manufacturing workflows
  - Distribution workflows

Be forewarned, however. You must validate that department personnel actually follow the workflows and procedures as outlined in these documents before accepting them as an accurate portrayal of real business practices.

In contrast, other operating units run entirely on oral history. The only way to determine how things are done is to interview the key stakeholders at all levels within that business unit. Other sources of information to consider (if they are available), include departmental histories, operating manuals, prior IT organization research, meeting minutes and systems documentation concerning the IT systems already in place, customer and employee survey information, and the written and verbal musings of operating personnel. Do not rely entirely on the views of management. Managers will more often than not tell you how things should work rather than how they actually do work. Scrutinize whatever information the project team receives. Take nothing at face value, and ensure that you get a balanced viewpoint. All in all, be critical of what you read, hear, and learn.

Unfortunately, there may be a disconnect between what your sponsor and working clients believe must happen as a consequence of your IT project and what in truth is required. You need the support of management to proceed, so do not be antagonistic. Listen respectfully, but reserve judgment. Remember that your purpose is to collect objective business process requirements and to use these data points to move both the business and the IT sides of the house toward consensus about the scope and nature of the IT project at hand. By employing a simple lens and asking open questions, the team's business analysts should obtain the spectrum of information required for further action. To this end, your process of data collection and analysis must be deliberate, methodical, and as free of prejudice as possible.

Once the project team has identified its targets for analysis, it should obtain the backing of the sponsor and working clients to proceed. The true test of that support will be evidenced by the quality and quantity of time that line-of-business participants spend with you and your team. Sometimes, it makes sense to formally charter the data-gathering effort, defining these elements:

- Roles and responsibilities of business and technical personnel
- Scope of information to be collected and reviewed
- Desired data collection and analysis outcomes
- Timelines for the completion of tasks (a mini-project plan)
- Process metrics

For its initial set of activities, the project team would then conduct a quick, high-level mapping of the business process to derive a general knowledge of requirements. To that end, the team must explore both the business and technical sides of each business process by examining the following components:

- Current state of business process knowledge components:
  - Project charters

- Contracts
- Standing business requirements
- Project plans
- Resource plans
- Budgets
- ROI analysis
- Process models and maps
■ Current state of technical process knowledge components:
  - Operations scenarios
  - Operations documentation
  - Test plans and scripts
  - Performance metrics
  - Technical documentation
  - Change orders

As a next step, the team would complete an information system asset inventory and gap analysis, identifying the gaps in the current IT offering and any additional attributes desired of the enhanced or replacement system. Armed with this information, the team can proceed to identify those aspects of the process and its business functionality that require more in-depth research and understanding. As a final phase in the business requirements gathering effort, the team will construct a formal set of business requirements to be reviewed in detail by appropriate members of the business team and ultimately signed by the sponsor and the working clients.*

In concluding this work, the project team will have the business requirements that it needs to move from analysis to design. But getting to customer signoff can be tricky and is almost always laborious. Although your sponsor may want the project done, he or she cannot understand why all this time must be spent in discovery. Unfortunately, you will find that some of your customers do not want to be held accountable and are therefore annoyed by the rigor of the requirements process. Similarly, you and your team do not want to be accused of analysis paralysis by either business or IT management. Yet to blunder ahead without adequate attention to business requirements can only end badly when you turn to

---

* During the course of data gathering, the project team will identify key business process participants, who, due to their experience, knowledge, tenure, and veracity, have won the respect of the team. The project director must find ways to ensure that these folks are engaged in the process of project requirements development and review. When management is respectful of these employees and their opinions, bringing them into the formal process makes sense. But when management is clearly uncomfortable with their involvement, the project team should find ways to involve them informally. It would be a mistake to discount the views of these employees merely because they are out of favor with management.

IT solution development and delivery. To balance these conflicting demands, use a thorough methodology that quickly lays out the entire business case in sufficient detail. You may then return and drill on those components that require more in-depth scrutiny. The process mapping technique demonstrated in this chapter will readily gather most, if not all, the information required to generate a satisfactory framework and blueprints for a formal IT solution design.

BUSINESS PROCESS MAPPING

The business process mapping methodology encompasses seven key sets of deliverables:

- Process decomposition, including:
  - Brief process definition and overview
  - List of process assumptions and operating principles
  - Process workflow
  - List of process inputs, outputs, and associated customer deliverables
  - Map or checklist of the information technologies that enable, and automate the process
- Roles and responsibilities matrix
- Set of approval rules governing the process
- List of access rights (to data, systems, etc.) for those operating within the process
- Performance metrics that measure process outcomes
- Process templates, frameworks, standards, and tools (i.e., the process' reusable components)
- So-called knowledge library of explicit process knowledge, such as scenarios, case studies, best-in-class examples, and linkage to internal and external supporting information resources

Together, these process mapping elements provide a complete picture of the business's current or desired workflows and its underlying IT requirements. It is assumed that one or more business analysts from the PMO will perform much of this work on behalf of the IT project team. The remainder of this chapter examines these mapping elements in some detail, employing the solution selling process within the HG & Co. consulting firm as a case study example. But before we proceed, the author would like to make a brief case for employing a business process mapping methodology.

First and foremost, business process mapping focuses on the underlying business value(s) of the process under consideration. If the enterprise is to invest its human and financial capital in an IT-enabled business process, the value proposition for that investment must be clear and cogent from the outset. In making it so, proponents of the undertaking will enjoy the ongoing support of key executives, allowing the project team to devote its energies to project execution. Mapping also helps to define business priorities and to identify those particular business unit performance problems around scrap and rework, poor communication, and an inability to leverage knowledge and IT assets to achieve the enterprise's goals and objectives.

Mapping also creates a holistic view of the current state of realities around business process delivery. From this, mapping leads to gap analysis and the definition of a desired state that in turn clarifies the respective positioning of IT enablement, work process collaboration, and the sharing and leveraging of enterprise resources. Of course, the devil is in the details. So let us consider the elements of business process mapping.*

PROCESS DECOMPOSITION

As the first set of mapping activities, process decomposition reveals the rationale behind each key business process and its underlying assumptions and operating principles. It requires a clear and complete high-level map or workflow of how the enterprise executes that process, citing all its essential business rules, process components, and deliverables. As a complement to this map, a business analyst prepares a table that defines the inputs, outputs, and deliverables for each process step. This element documents the value chain of the process under analysis. Lastly, as part of process decomposition, the team maps major process components to their associated enabling technologies. In our example, solution selling within the HG & Co. consulting firm, the business process may be defined as follows:

- Team-based, problem-focused selling process that proactively identifies business opportunities and presents the value proposition for addressing those opportunities to targeted, qualified clients
- Holistic process that encompasses diagnosing, proposing, and closing a business proposition that solves a client business problem
- Selling solutions that are tailored and customized to the particular needs of a given client, even though the actual deliverables may be cast from standardized components

---

* For an electronic version of the author's mapping tool, see *The Hands-On Project Office*, http://www.crcpress.com/e_products/downloads/download.asp?cat_no=AU1991, chpt6~1~process map~template. Also included are two completed mappings: one for the selling solutions example cited in this chapter (see chpt6~2~process map~selling~example) and the other a related mapping for service delivery, again featuring HG & Co. (see chpt6~3~process map~execution~example).

- Intense collaboration to derive the right result (often a cross-organization, team-sell approach is required)
- *Not* a product sheet and a price but a process that draws on unique competencies and knowledge bases to create a complete solution for the client

Note that this definition is concise and explicit, providing all interested parties with a clear sense of what the firm means by solution selling. Note also the references to knowledge bases, implying an underlying set of IT-enabled components to the process. The solution selling's assumptions and operating principles further clarify the working definition of the process:

- Solution selling is an iterative process
- Negotiating and communicating with the client throughout the process is key; antennas up for client pain, restraints, preferences, etc.
- Price should not be discussed for the first time with client in the proposal; it should be addressed as early as possible and certainly during scoping and scope agreement
- Consultants must involve sales as they contact clients or prospective clients, and they must record these efforts in the appropriate relationship log in either the salesforce automation system or the professional services administration system
- Whoever initiates a solution selling instance must get others involved as early as appropriate to ensure the process's success
- Process must scale with the business
- Where possible and appropriate (e.g., for off-the-shelf products and services), payment in full should be required up front
- No new work will take place without a credit check and the clearing of past-due balances (accounts)

Here, the firm's process owners have clearly articulated norms of behavior and performance standards, if not desired process outcomes. From the IT project team's perspective, this information begs the question, "How can IT enable and empower this process?" To answer this question, the IT team must identify and map the entire process in some detail, then examine each process activity for investment opportunities. As a start to formal mapping, simply make a list of all key process steps in sequential order. Do not worry about the rigor of your sequencing or leaving out details. Your list is merely an aid in drawing the process map, which will accommodate concurrent activities. Its execution will also solicit from your working client(s) and the project team process steps that you may have neglected. The list for solution selling process steps runs as follows:

Collecting and Capturing Business Requirements for IT Projects ■ 157

- Target prospect
- Contact/interact with the client
- Identify/define opportunity
- Validate and agree on problem/opportunity (the firm and the client)
- Qualify client
- Identify buying criteria
- Scope potential assignment
- Agree on scope with client
- Set price (start negotiating with client)
- Approve/sign off proposal to client by senior management
- Prepare and present proposal (continue negotiating with client)
- Prepare and present contract (finalize negotiations with client)
- Obtain purchase order/PO number from client before the start of the assignment
- Kick off project
- Track interim process results
- Track process outcomes
- Report on process results to appropriate project stakeholders

With this checklist in hand, the time has come to develop a business process map. See Exhibit 1 (where *BU* equals *business unit*).

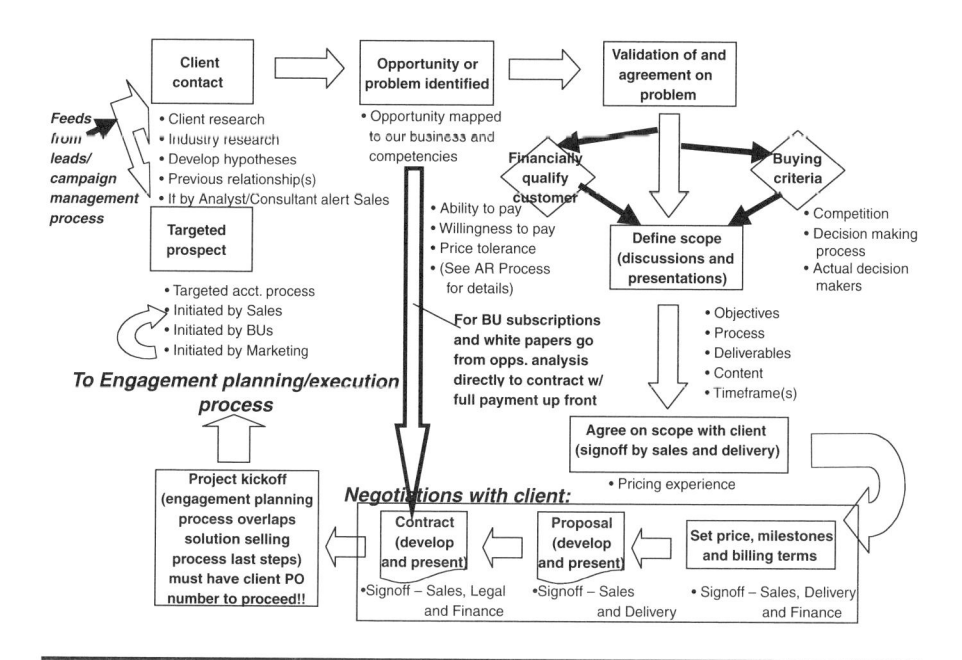

Exhibit 1   A Business Process Map for the HG & Co. Selling Solution

This example of a process workflow is not particularly detailed, nor does it need to be. Other process mapping elements complement the actual flow diagram and fill in the particulars. Nevertheless, the diagram does capture primary process steps, decision trees, concise roles and responsibilities, summary business rules, and desired process outcomes. With this blueprint in hand, the project team and its working client partners will next create a simple matrix that lists the following:

- Each process step
- Inputs and outputs of that step
- Party or parties responsible for that step's execution and deliverables
- Customer/process deliverables emerging from that step

This exercise makes explicit what is perhaps implied in the workflow. Furthermore, it formally sets down the more incremental elements of the process, as well as the desired outcomes and the responsible parties. In the case of our solution selling example, the matrix appears as in Exhibit 2.

To conclude mapping process decomposition activities, capture the current state of those IT systems that enable the business process being mapped. For example, in the case of solution selling, the HG & Co. consulting firm employs a lead/sales campaign management system, a customer relationship management (CRM) system, and a professional services administration system (for engagement management and delivery, and for project cost accounting). See Exhibit 3.

The importance of this modest step may not be apparent but in acknowledging the information systems that currently support the business process, the IT team identifies both potential sources of knowledge content and potential technology platforms for constructing an enhanced or replacement IT solution. For example, HG & Co. Consulting's CRM system tracks sales cycle data. With this information, the IT team can enable an information service that links new sales personnel with sales veterans who possess the requisite expertise. Similarly, the professional services system already houses project engagement knowledge that could be shared with new service teams faced with similar assignments. In total, the elements of process decomposition provide the IT project team with a rich understanding of the business process, its information management requirements, and the knowledge artifacts that it generates for sharing and reuse.

THE ROLES AND RESPONSIBILITIES MATRIX

As part of the process decomposition effort just described, a business analyst has already prepared an input/output/deliverables matrix. Next, the team should map clearly defined and specific roles and responsibilities

Exhibit 2 The Components of Process Decomposition

Process Step	Inputs/Outputs Described	Role Owners	Customer/Process Deliverables
Target Prospect	Inputs: market/prospect research and analysis Outputs: targets agreed to	Analysts, Sales, BUs	List of qualified targets
Contact/interact with the client	Inputs: qualified targets Outputs: prospects approached	Consultants, account managers	Approach to specific target
Identify/define opportunity	Inputs: industry, market, prospect information Outputs: specific opportunities defined; if an off-the-shelf product, go directly to contract	Sales, consultants	Sufficient information to craft an approach for prospect; contract
Validate/agree on problem/opportunity	Inputs: opportunities; ideas presented Outputs: consensus on focus for sales and delivery teams	Sales, delivery, client	Consensus with client on opportunity under consideration
Qualify client	Inputs: credit checks; client financials and information Outputs: green light on client's ability and willingness to pay; determination of client's price tolerance	Sales, finance	Determine client's ability and willingness to pay; price tolerance
Identify buying criteria	Inputs: client information; competitive intelligence Outputs: alignment of the firm's selling process with prospect's buying process	Sales	Determine how decision is to be made and by whom; competition

(continued)

Exnibit 2 (continued) The Components of Process Decomposition

Process Step	Inputs/Outputs Described	Role Owners	Customer/Process Deliverables
Scope potential assignment	Inputs: client information; process intelligence Outputs: scope document	Sales, delivery	Scope document
Agree on scope	Inputs: scope document Outputs: consensus on scope	Sales, delivery, client	Consensus with client on scope
Set price	Inputs: scope document; pricing experience Outputs: pricing for proposal	Sales, delivery	Based on scope, price determined
Prepare and present proposal	Inputs: scope document and pricing decision; resources availability Outputs: proposal	Sales, delivery	Proposal delivered
Prepare and present contract	Inputs: proposal and negotiations Outputs: contract	Sales, delivery	Contract delivered
Kick off project	Inputs: contract Outputs: engagement and delivery plans	Engagement management team	Project initiated

Collecting and Capturing Business Requirements for IT Projects ■ 161

Services required:

Targeting prospects	Tracking sales cycle process	Proposal management	Templates, knowledge bases, tools, et al.,	Contract management	Engagement management

Enabling technologies:

I/T key:

Leads/campaign management	SalesLogix	Niku

Exhibit 3   A Map of Business Process IT Enablement

for each process step. This activity is essential to ensuring the commitment of each process participant to his or her role within the project team's proposed model for enhanced or reengineered business process delivery. For the IT project team, this information is also essential in identifying those who will create or provide knowledge components for others to use within the process's value chain. In addition, this mapping element often documents the sequencing of component handoffs within the larger business process.

It should come as no surprise that some of those involved in process delivery are unwilling to acknowledge their actual roles and responsibilities. The mapping process helps these folks face up to their own accountability. For some, the information in this matrix may come as a real surprise. Whatever the particular circumstances within your organization, it is essential that all business process stakeholders recognize and assume responsibility for their process roles. By achieving this aim, the IT project team will find it easier to build these assignments into new IT systems and IT-enabled business processes, and to define more clearly task handoffs, labor-saving opportunities, and so forth. The actual execution of a roles and responsibilities matrix is demonstrated in the solution selling example. See Exhibit 4.

Note that the matrix itself should follow the same order as the process workflow map and should use clear, unambiguous language in the definition of assignments and outcomes. Do not be surprised if this step, which some might view as a formality, becomes a major point of contention. Stakeholders do not always want to be reminded of their responsibilities, nor do they wish to have their accountability documented as part of a business process mapping exercise. Nevertheless, for reasons that are obvious, this work must get done. Furthermore, the time spent to discover

Exhibit 4   A Roles and Responsibilities Matrix

Role	Responsibilities	Customer/Process Deliverables
Consultants, sales, Business Units	Collaborate on defining targets; BU and Sales leadership to provide direction, build consensus, determine priorities	List of qualified targets
Consultants, account managers	Appropriate employee/team to approach the targeted prospect and to gather information — both in the marketplace and from the prospect	Approach to specific target
Sales, consultants	Designated solution selling team to distill information into a specific set of opportunity scenarios and options	Sufficient information to craft prospect approach
Sales, delivery, client	Designated solution selling team to meet with prospect to explore a specific set of opportunity scenarios and options and to agree upon a focus and priorities, level of interest	Consensus with client on opportunity under consideration
Sales, finance	Due diligence to ensure that the firm follows the prospect's buying process, that we are talking to the right people, and that we know who our competition is	Determine how decision is to be made and by whom; competition
Sales, delivery	Prepare a cogent scoping document upon which the project team can deliver if engaged	Scope document
Sales, delivery, client	Reach consensus and commit in terms of client's needs, firm's capacity, etc.	Consensus with client on scope
Sales, delivery	Due diligence to set price based on past delivery and pricing experience	Based on scope, price determined
Sales, delivery	Prepare, deliver, and present proposal; reach agreement	Proposal delivered
Sales, delivery (?)	Prepare, deliver, and present contract; get signoff	Contract delivered
Engagement management team	Ensure there is clarity and commitment in the handoff from the sales team to the engagement team executing the contract	Project initiated

the current state of the process and to win overall agreement on the desired state is essential to subsequent IT system design and development. Bear in mind that your business analysts will not have the clout to accomplish consensus building. They are there to gather the facts. The more dynamic and interpersonal task of reaching a common view of the replacement business process must be left to the project director in partnership with his or her project sponsor and working clients.

PROCESS RULES

Along with the business process mapping and the defining of participant roles and responsibilities, the project's analyst should capture all of the business rules underlying the now documented process. Just as workflow steps, task assignments, and customer deliverables are all part of the framework that governs a business process, each process has its decision points and associated approval workflows. If automated systems are to be employed, approval rules must be explicit so that the rule-based engines in these systems may be appropriately programmed for use within the process. In our solution selling example, the process approval rules include the following:

- Sales and BU leadership approve targeted prospects
- Sales leadership approves contact person and process for initial prospect contact
- Consultants must inform sales management when they contact an existing client or targeted prospect concerning a new opportunity
- Assigned solution selling team (made up of sales and consultants as the case may be) develops the initial opportunity analysis
- Appropriate sales/delivery team deals with the prospect on the problem statement
- Sales and finance management qualify the prospect
- Sales, delivery, and finance management approve pricing and the proposal going to the prospect
- Sales, finance management, and legal sign off on the contract before it goes to the prospect
- Delivery drives the kickoff process with the support of sales

Such rules may direct the flow of information from one process stakeholder to the next and hence influence the overall design of the process's underlying IT platform. Similarly, when a process calls for the calculation of values or other data manipulations, there could be any number of business rules associated with these functions within the process.

Yet another consideration in process mapping and subsequent design is the rules governing stakeholder access. Depending on the sensitivity of and legal restrictions placed upon information employed by a particular process, the enterprise must exercise control over employee and customer access rights. In a fairly open corporate culture, where the content of a particular knowledge base is generally free of legal and operational concerns, access rights and individual user profiles may be broadly defined. In many instances, however, rigorous access rules are required. During the mapping process, the IT project team will collect access requirements and build standard user access profiles as need be. In the case of the HG & Co. consulting firm, solution selling data, though highly confidential to the outside world, is readily shared within the organization. The firm therefore employs rather modest access rules:

- Sales and delivery teams will have full access to all enabling systems, including the sales force and professional services administration systems as needed
- Sales and delivery teams will have full access to all data, templates, tools, etc., via the intranet and the professional services administration system portal
- Modeling templates will retain individual employee confidentiality concerning salaries, etc., to the extent possible, while providing sales and delivery teams with the information they require to scope and commit to projects

Note the clear and unambiguous implications of these rules for the eventual design of an enabling IT platform. Your system's access rules may be much more complicated, especially if they involve content delivery to customers who reside outside the perimeter of your information security firewall. In most instances, you will want to involve both those with fiduciary responsibility for the data and the enterprise's information security officer in deliberations concerning access controls.

PERFORMANCE METRICS

Every key business process employs a series of largely quantifiable performance metrics that measure both process effort (execution) and results (business objectives achieved) in a meaningful fashion. Surprisingly, the IT project team may discover that the process under examination either lacks performance metrics or employs surrogates that measure activity rather than results. As part of process mapping, standards and mechanisms of measurement should be identified — or

Collecting and Capturing Business Requirements for IT Projects ▪ 165

created from whole cloth — and documented. Whenever possible, meaningful, results-oriented metrics should be deployed. As appropriate, actual measurement should take place through the automated system(s) that enable the business process in question. But even automated systems may not lend themselves to meaningful measures of success, necessitating creative solutions in which other IT operating units may play a supportive role. For example, if the envisioned IT solution involves the delivery of Web services, the project team may work with its Web technology platform colleagues to install a monitoring tool, like WebTrends, that will provide meaningful data about how customers use the newly implemented offering.

As a case in point, here are HG & Co. Consulting's solution selling performance metrics. Not all of these measures are hard (and some reflect wishful thinking).

- Close rate metrics:
  - Dollars closed versus dollars in pipeline
  - By project and in total: dollars proposed versus dollars committed to in the contract (i.e., contract value)
  - Trends (e.g., types of products or services sold, types of customers)
- Sales volume metrics:
  - Dollars scoped per client opportunity (average-size deal)
  - Proposed opportunities versus closed opportunities (closure rate)
  - Total dollar volume
- Cost of sale metrics:
  - Time charged by all participants in solution selling process versus total dollar volume
- Sales cycle time metrics within each of the following cycle elements: opportunity definition, scoping, proposal completion, contract completion:
  - Duration of process from initiation to close
  - Duration of key process elements
- Project hurdle rate metrics:
  - Price of proposed deliverables per hour of labor
- Wins metrics:
  - Success stories, qualitative information

As project personnel with process engineering expertise, the team's business analysts will help stakeholders define appropriate measures and assist the IT development team in building mechanisms and processes within the enterprise's IT environment that both capture those measures over time and provide longitudinal data for the analysis of business process results.

PROCESS TEMPLATES AND TOOLS

During mapping, the IT project team will discover that most explicit process knowledge and best practices are captured in the templates and tools employed as part of that process. Typically, the team's business analyst identifies those templates and tools already in place, as well as those required to streamline and enable the process further. As part of IT platform design, these components are linked to their respective process steps for ease of reference, retrieval, and reuse. Do not be surprised if mature business processes possess numerous duplicate templates, style sheets, and the like. As part of its value proposition, the IT project team will identify how best to rationalize this body of work, distilling and promoting best practice and enterprisewide standards. In many instances, manual processes with various associated paper forms will be replaced by more uniform, paperless, online workflows.

Here is what the HG & Co. consulting firm's IT project team uncovered by way of existing solution selling process templates and tools:

- Opportunity/problem assessment
  - Product/service diagnostic tools
  - Assessment templates
  - Previous working models and sizings
  - Case histories and related deliverables
- Client qualification
  - Qualification questions
  - Qualification techniques
- Buying criteria
  - Buying questions
  - Buying techniques
- Project scoping
  - Templates
  - Scoping questions
  - Client/prospect information
  - Pricing guidelines
  - Project/delivery methodologies
- Pricing
  - Models, rate sheets
  - Work plans
- Proposal templates
- Contract templates
- Kickoff template(s)

In the course of subsequent analysis, some of these components will be retained and others discarded, redesigned, or replaced.

In building a new IT-enabled platform, the payoff will come in linking standardized templates and tools to the business process itself; when a stakeholder comes to a given process task, the appropriate templates and so forth are automatically presented for use. When multiple options exist, the user should be directed to the correct form; when no tools are in place, the IT business analyst will collaborate with the working client(s) to create standards. There is nothing particularly difficult about this IT task. It merely requires a thorough understanding of the business process and its stakeholders, as derived from the mapping effort itself, and a disciplined, objective sense of best practice in building automated workflows and decision trees.

The greatest challenge will come in selling new practices to those who are wedded to old ways. By aligning performance metrics with these new practices and demonstrating positive outcomes in line with these measures, the project team well be in a strong position to sell the value of these process changes to stakeholders. As is always the case with any business process transformation, however, getting the actual participants to embrace new standards and practices is the real challenge. Here, it is best to remember why your team has taken such pains to involve your project's sponsor and working clients in your efforts. It is up to the line-of-business leadership to bring its team along.

BUILDING A FINAL PICTURE OF THE IT SOLUTION FOR THE CUSTOMER

Process templates do not tell the entire story. Your business customers will also benefit from seeing examples of completed templates, case studies of how to employ newly IT-enabled process tools, illustrations of the new business process's end products, and so forth. Some of these components or artifacts already exist and need only be appended to the IT process reengineering documentation. Others may exist, however, only as tacit (i.e., undocumented) knowledge, garnered as part of the business analyst's labors and interactions with business unit personnel. Here, the enterprise will require some incentive that encourages in-house experts to convert their experiences to sharable knowledge assets. The IT project team can provide support for these individual efforts, as well as the linkages between their outcomes (e.g., the discrete artifacts that they generate) and the overall business process/IT framework.

Even in the best of all possible worlds, only a fraction of useful tacit knowledge will be transferred to more explicit forms. Few organizations have the time or resources to indulge in oral history projects. Instead, the IT project team may, over time, develop IT-enabled exchanges and forums to link information seekers more easily with subject-matter experts. All of these process artifacts and experiences serve as a business process knowledge library to be used by less experienced personnel. In the instance of the solution selling case study, this library of components includes the following:

- Process templates
- Market intelligence
- Selling solution case studies by customer type and product
- Project scoping tools and illustrations
- Proposal library
- Solution selling process discussion forum
- Directory of solution selling experts (e-mail enabled)
- Internet training courses
- Related online publications

As with the other elements of the process mapping effort, this knowledge library component affords valuable input into the design and development of an enabling IT platform.

Finally, to flesh out the actual design of the IT solution, consider a simple approach to translating the business mapping process findings into a more graphical representation of next steps. Exhibit 5 integrates business process steps with roles and responsibilities and the associated enabling technologies. The graphic presents the envisioned IT-enabled business solution as a series of layers. Beginning with a Web interface for customers and enterprise personnel alike, I outline customer-facing, service delivery, and infrastructure layers. Each layer defines the IT project director, external IT partners, internal IT liaisons or support personnel, and the associated business process owner(s). The most complex layer, the service delivery layer, also identifies, by process subcategory, the power users of each IT system. Lastly, the illustration depicts the IT products that must be brought together to deliver the IT solution called for as an outcome of the business process mapping exercise.

CLOSING COMMENTS

These pages have discussed methods for the detailed documentation of the business process analysis and requirements gathering effort within a larger IT project. Although there are many reasons why an organization might turn to process mapping, the purpose here is to expose its benefits as a tool set in facilitating IT project analysis and design. With the data derived from this collaborative and consensus-driven process, the IT project team is in a much better position to win support for the reengi-

Collecting and Capturing Business Requirements for IT Projects ■ 169

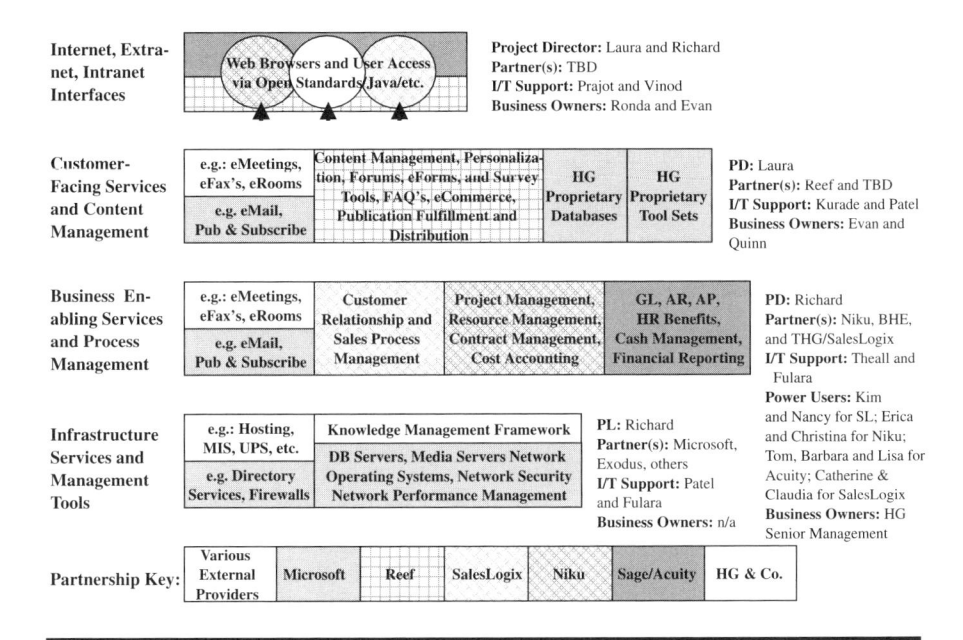

Internet, Extranet, Intranet Interfaces	Web Browsers and User Access via Open Standards/Java/etc.					**Project Director:** Laura and Richard **Partner(s):** TBD **I/T Support:** Prajot and Vinod **Business Owners:** Ronda and Evan
Customer-Facing Services and Content Management	e.g.: eMeetings, eFax's, eRooms e.g. eMail, Pub & Subscribe	Content Management, Personalization, Forums, eForms, and Survey Tools, FAQ's, eCommerce, Publication Fulfillment and Distribution		HG Proprietary Databases	HG Proprietary Tool Sets	**PD:** Laura **Partner(s):** Reef and TBD **I/T Support:** Kurade and Patel **Business Owners:** Evan and Quinn
Business Enabling Services and Process Management	e.g.: eMeetings, eFax's, eRooms e.g. eMail, Pub & Subscribe	Customer Relationship and Sales Process Management	Project Management, Resource Management, Contract Management, Cost Accounting	GL, AR, AP, HR Benefits, Cash Management, Financial Reporting		**PD:** Richard **Partner(s):** Niku, BHE, and THG/SalesLogix **I/T Support:** Theall and Fulara **Power Users:** Kim and Nancy for SL; Erica and Christina for Niku; Tom, Barbara and Lisa for Acuity; Catherine & Claudia for SalesLogix **Business Owners:** HG Senior Management
Infrastructure Services and Management Tools	e.g.: Hosting, MIS, UPS, etc. e.g. Directory Services, Firewalls	Knowledge Management Framework DB Servers, Media Servers Network Operating Systems, Network Security Network Performance Management				**PL:** Richard **Partner(s):** Microsoft, Exodus, others **I/T Support:** Patel and Fulara **Business Owners:** n/a
Partnership Key:	Various External Providers	Microsoft	Reef	SalesLogix	Niku	Sage/Acuity	HG & Co.

Exhibit 5 An IT Road Map for Implementing Business Process Change

neering of business practices and the investment in information technology required to convert process findings into implementable project plans and system development efforts.

To summarize, the steps in this chapter's business processing mapping methodology include the following:

- Process decomposition — captures the detailed workings of the process, information flows, process outcomes, and so forth. This includes process definition and overview, assumptions and operating principles, workflows, inputs, outputs and deliverables, and associated enabling technologies.
- Roles and responsibilities matrix — documents individual stakeholder commitments within the process, as well as opportunities for automated information handoffs.
- Process rules — define approval rights and access rights within the process and, by implication, the knowledge requirements of stakeholders in decision-making roles.
- Performance metrics — include the key measures of process delivery and customer satisfaction that the IT process captures, shares, and leverages as part of its value proposition to the enterprise.
- Process templates — include standardized knowledge assets — like templates, models, and tools — that are typically standardized, main-

tained, and distributed via an IT platform and, in part, through the efforts of the IT project team.
- Knowledge library — encompasses nonstandard knowledge assets: either formal artifacts (i.e., explicit knowledge) or services for accessing or exchanging tacit knowledge.

Once completed, an enterprise's process maps will provide the IT project team with ample direction to construct an impactful automated platform for the sponsoring business unit. Success in process mapping calls for special skills that align with the competencies of a business analyst (see Chapter 2). The virtue of a PMO in this regard is that it brings together a core group of analysts who can collaborate on process mapping assignments and share their experiences across the team. Although there is definitely a benefit to embedding analysts within business units so that they can master the ways in which those organizations work, it is of even greater value to retain them as objective third parties. This allows them to bring a certain freshness of perspective to each new project, even as they draw upon the collective experiences of their PMO colleagues and the knowledge assets created and shared among team members.

Chapter 7 explores in more detail how knowledge management, as championed by the PMO, may serve the greater IT organization.

7

MANAGING LESSONS LEARNED — THE REUSE AND REPURPOSING OF IT ORGANIZATIONAL KNOWLEDGE: A CASE STUDY

INTRODUCTION

As a process, knowledge management (KM) is as old as humanity's need to record, retain, and share information. Its advent preceded the invention of the computer, and indeed, the use of paper as a medium for the exchange of knowledge. This chapter does not consider the long and circuitous history of KM, nor does it focus on KM as a theory, a philosophy, or an academic discipline. Rather, the primary purpose of this chapter is to demonstrate the utility of applying knowledge management within an IT organization and the integrative role to be played by the PMO in the establishment and ongoing delivery of KM services. For the purposes of this discussion, I draw upon my recent experience developing a KM platform for the information services (IS) division of Northeastern University. This case study illustrates the benefits in leveraging KM to strengthen the performance of the IT organization.

Like most IT organizations running within a larger enterprise, Northeastern University's IS division operates from a number of locations, handles a multitude of products and services, and serves a diverse customer base. Furthermore, in striving to meet the expectations of its constituents, the IS division must overcome many of the same challenges faced by IT organizations elsewhere, including the growing complexities and interdependence of IT applications, the need for better systems integration, the shortage of technology expertise, the paucity of available financial resources — especially for technology currency — the customer's demands for speedier time to market, the growing concerns over disaster recovery and business continuity planning, and so forth. For its part, the university's leadership is demanding more return in terms of measurable results from the institution's investment in computer hardware and systems.

Early in 2001, driven largely by the demand to do more and better with less, Northeastern's IS division established the institution's first-ever project management office (PMO): IS Enterprise Operations (EO). In late 2001, EO launched an effort whose primary objective was to create a Web-enabled platform that fosters collaboration through the leverage and reuse of IS knowledge. As conceived by the division's KM team,[*] this Web portal brings together for easy access a vast body of explicit knowledge derived from the project plans, functional requirements, and technical specifications to product white papers, training materials, and customer feedback survey data of IS working groups.[**] In addition, the portal links the IS team to related information-sharing and communication services, such as (Lotus) Domino TeamRooms and discussion databases, electronic mail, instant messaging, Remedy (the division's problem resolution/trouble ticket tracking system), and staff alerts and news items. Last but not least, the portal captures staff biographical data and expertise levels so that individuals may mine the division's tacit (i.e., unrecorded) knowledge through the facilitated identification of local experts.

Creating a knowledge-sharing platform for an IT organization poses its own special challenges. This chapter explores the steps involved in designing, building, and maintaining a portal environment. Also, it includes a discussion of the PMO's approach to knowledge management in service of the greater IT organization and comments on the underlying technology enablement of the portal itself. Last but not least, this chapter reviews the challenges faced by any PMO in maintaining a KM platform whose purpose is to improve IT service and project delivery through virtual knowledge sharing and facilitated collaboration.

---

[*] Although Northeastern University offers programs to certify knowledge managers, the IS division does not formally employ knowledge management personnel. Rather, as part of its mandate, Enterprise Operations (NEU's PMO) took on a KM role as this concerns the knowledge and learning processes within the university's IT organization.

[**] In this context, a portal refers to an Internet site that aggregates content and services for the so-called one-stop-shopping convenience of its users. In the case of an IS organization, such a portal would include alerts and news of particular relevance to the team, working documents, template libraries, links to IS tools, and so forth.

THE WHATS, WHYS, AND WHEREFORES OF KM

Within an enterprise, KM is that process which identifies and brings to bear relevant internal (i.e., from within the enterprise) and external (i.e., from outside the enterprise) information so as to inform action. In other words, information becomes knowledge as it enables and empowers an enterprise's work force, known in this context as its knowledge workers. The information components of any KM process come in one of two flavors:

- Explicit knowledge — structured or documented knowledge in the form of written reports, computer databases, audio and video tapes, and so forth
- Tacit knowledge — undocumented expertise walking around in the heads of the enterprise's knowledge workers or external third party subject or process experts

The task of a KM process is to organize and disseminate explicit knowledge, as well as to bring together knowledge workers and the appropriate information — explicit and tacit — required for their assignments. The greatest challenge to those charged with KM is the targeting and conversion of tacit to explicit knowledge and its timely delivery to those in need.

Before the Internet, many business leaders viewed KM simply as an expensive and somewhat frivolous undertaking. But as it has become clear that well informed employees and IT-enabled business processes positively impact enterprise performance, management resistance to KM has softened. Furthermore, the arrival of the ubiquitous Internet has provided organizations with a relatively low-cost platform for information gathering and sharing. In brief, the value proposition for KM is as follows: KM optimizes the value of what workers know in delivering value to customers, and hence KM serves as an enabler of performance excellence.

Knowledge management is all about enabling, empowering, directing, and energizing the work force in fast-paced, evolving, and increasingly virtual enterprises. A partial list of the tangible benefits from such a KM program might very well include the following:

- Better understanding of the marketplace and customer needs
- More effective sales process as measured in terms of customer retention
- Faster and higher quality product/service time-to-market
- Reduced operating costs and overhead
- Reduced new venture/current operating exposure

- Higher level of innovation
- Broad-based adoption of industry and process best practices
- Higher employee retention

Of course, the vendors of IT products and services make the same benefit claims. The author has no intention of overselling KM or its underlying technologies. Knowledge management may work for your organization, but getting there and sustaining the effort is not easy. This chapter and case study will tie a field-tested KM enablement process to the key drivers of success for an IT organization, illustrating the possibilities as well as the barrier to its successful application. Although the focus of this case study is an institution of higher education, the business issues and IT challenges faced by Northeastern are common to any multi-location enterprise with diverse lines of business serving a heterogeneous consumer population.

GETTING STARTED: INTRODUCING THE CASE STUDY

Northeastern University (NEU) is a large, urban university recognized nationally and internationally for its commitment to applied learning and cooperative educational programs. Its main campus is located in the Back Bay of Boston, Massachusetts; NEU also delivers programs through several satellite campuses in and around Boston. The university includes schools of engineering, health sciences, computer science, business, law, criminal justice, and the arts and sciences. The NEU community serves over 30,000 students — including undergraduate, graduate, and nontraditional students — through the efforts of more than 4,000 faculty and administrative personnel. In total, the greater university campus includes more than 80 buildings, operating largely off a common, state-of-the-art optical fiber network backbone and running a vast array of IT platforms and products. The diverse nature of the Northeastern's investment in information technologies is typical of what one finds in higher education, where program offerings drive diverse IT choices and, therefore, the maintenance of a rather heterogeneous IT environment.

To manage this complexity, NEU funds a centralized IS division whose responsibilities encompass the ongoing support and maintenance of the campus's IT infrastructure of networked hardware and software, all university administrative and business systems, and all common campus IT services, such as the Internet, electronic mail (Lotus Notes), desktop automation tools (e.g., Microsoft Office), and group collaboration tools (e.g., Domino TeamRooms, Domino Discussion Databases, Lotus Instant Messaging, and Notes calendaring and scheduling). In addition, IS operates a number of large, public computer labs for the student body, a residence hall computer network on the Boston campus with approximately 8,000 network nodes, a walk-up help desk, a call center, campus-wide multimedia and audio/visual services, and a body of end-user documentation and training programs in support of core university IT offerings. Finally, IS works closely with the university's various colleges and schools, each of which sponsors its own IT organizations, to ensure that the program-driven technologies within these operating units integrate with the university's overall IT infrastructure.

This range of activities poses obvious challenges for IS. Clearly, the division manages services and executes projects that impact the entire NEU community. To that end, effective customer relationship management and communications, as well as successful delivery of products and services, are critical to the success and credibility of IS. Because most IS tasks require the complex coordination of working clients, various IS teams, and an array of external technology partner providers, coordinating all the moving parts in these processes is always difficult. So, too, is the timely communication of project and service statuses, issues, best practices, and lessons learned. Finally, within NEU's geographically dispersed work environment and where the residential population requires something approaching 24/7 service availability, IS personnel must have ready access to business and technical information and expertise across multiple disciplines. These challenges are further complicated by the fact that the IS organization itself is spread across campus limiting physical proximity to colleagues and making communicating and collaborating more difficult.

To meet customer expectations and achieve satisfactory results, the IS organization took steps in 2001 to break out of the mold of reactive, siloed service delivery. The leadership of Northeastern's IS division, Bob Weir (vice president and chief information officer) and Rick Mickool (executive director and chief technology officer) recognized the need for action and, as an initial step, established a new operating unit, enterprise operations — the NEU version of a PMO. The EO team's mandate included the following:

- Formalization of a planning process that would align IS resources and work with those deliverables of highest value to the enterprise[*]
- Creation and ongoing management of a project office that would promote best practices in project and service delivery

---

[*] For the details concerning this effort, see Richard M. Kesner, "Running Information Services as a Business: Managing IS Commitments within a Larger Organization," *Information Strategies Journal* 18, 4 (2002): 15–35.

- Quality assurance and release management (QA/RM) function that would ensure the effective and efficient release into production of IT solutions in line with customer requirements
- KM practice to capture, organize, and distribute explicit IS knowledge for the leverage and reuse of the IS division team.

As a package, EO services supported those on the IS division's front lines in their efforts to operate according to the best practices of higher education IT. In this capacity, EO was well positioned to provide KM services. The EO staff was directly involved in all major projects, providing both project management and business analysis support. In these capacities, EO team members facilitated or directly authored most of the key documentation associated with IS division planning and decision-making, with IT projects, and with the process-reengineering work between IS and its customers and within IS itself. Similarly, through the QA/RM functions, EO personnel created test scripts, product release schedules, and the monitoring and documenting of system bug fixes and source code management. Thus EO emerged as the primary source and keeper of IS best practices information. Given the pace and extensiveness of IS activities, however, it was not always possible for EO to share information or documents readily across IS. Furthermore, there was still no easy means for translating the expertise of individual IS experts into more generalized knowledge to be shared with colleagues.

To close this gap, Richard M. Kesner (then the director of EO) proposed, in late April 2001, the establishment of a formal knowledge store that would house all critical IS documentation and provide easy search and retrieval capabilities.[*] The envisioned service would facilitate many-to-many communication through a Web-enabled platform, allowing business process and technical subject experts to share their explicit (documented) and, to a more limited extent, their tacit (internalized) knowledge with their co-workers. Based on his prior experiences with KM,[**] the director of EO saw that this intranet solution, constructed as a knowledge portal, was the most cost-effective means of creating a proper environment for knowledge sharing.

---

[*] A knowledge store is a content repository in which an organization's documented processes and process outcomes are archived for easy retrieval. A cornerstone of any KM service, the store serves as a reference library, allowing the organization to leverage content for reuse and redeployment.

[**] As initially developed by KPMG and then modified by the author to complement his project delivery and knowledge management approaches, this model illustrates the various levels of service and the connections between a knowledge site and a PMO. See *The Hands-On Project Office*, http://www.crcpress.com/e_products/downloads/download.asp?cat_no=AU1991, chpt7~1~pm meets km~model.

To that end, he created a simple prototype, employing Microsoft PowerPoint. This model conveyed the basic functionality of a KM portal experience. He also created a Web asset inventory that offered further details concerning the functionality housed on particular portal Web pages and within site workflows. These illustrations convinced the IS vice president and executive director of the merits of the undertaking. They saw its value to the IS organization itself and as a prototype for similar KM efforts within other university business units. They therefore agreed to sponsor a development effort.

However, they also imposed certain limitations on the project. Because the knowledge portal project was not a key and immediate customer-driven IS deliverable, EO could not call upon the division's own, already stretched Web development resources for help in the project's creation. Through the limited funds provided, EO was charged instead with developing a solution in conjunction with some external technology partner provider. The scope of delivery would be bound by the limited funds available, but the timing of delivery was elastic, as was the definition of the initial functionality for the site. Thus, by the summer of 2001, the launch pad for the establishment of an IS knowledge portal project was in place.

The EO team, comprising the director and two business analysts — Beth Anne Dancause and Pam Marascia — (the division's core KM team), began to map a strategy for the project. From the outset, the team recognized the central role that a substantial knowledge store must play in the creation of a successful service offering. For this reason, and even before approval and funding were secured from executive management, EO personnel established a rudimentary document repository, employing a Lotus Notes TeamRoom. This so called IS TeamRoom was organized into a series of simple categories based on the various functional activities of the IS division, such as planning, benchmarking, service delivery management, operations reporting, and so forth.

To this the EO team added categories for major IS projects and a set of rules for the summary description of documents deposited in the repository. As IS team members created documents, an EO project manager or business analyst would deposit these under the appropriate TeamRoom subject heading (category). As needed, the EO KM team added new headings to maintain the team room's overall intellectual integrity. EO also ensured that documentation remained current and as comprehensive as possible. By late spring of 2001, all of IS had access to the team room, affording easy access to the knowledge store and other team room services, such as threaded discussions and publish–and-subscribe functionality.

The process of creating this interim knowledge store yielded a number of advantages to the greater knowledge portal project. First, it heightened the IS division's awareness of the broader interests and concerns of coworkers in the particulars of a given IS project or service delivery platform. Second, it instilled among those who actually created these knowledge assets a routine process for archiving and sharing documents. Third, it helped the EO KM project team to better discern patterns of usage among information resources and the associated cataloging requirements for document storage and retrieval. Lastly, the effort gave the KM team an excellent start in compiling the envisioned Web-based knowledge store. It was by now also clear, however, that as the team room grew in size and complexity, retrieval became more cumbersome. Indeed, within nine months, by the spring of 2002, IS staff began to complain that difficulties in finding specific IS TeamRoom documents discouraged use. Fortunately, by then the new KM portal was completed and ready to replace the IS TeamRoom. The balance of this chapter describes just how the portal team achieved this objective.

BUSINESS AND TECHNICAL REQUIREMENTS: ANALYSIS AND DESIGN

Like most projects launched over the past two years by NEU's IS division, the knowledge portal project followed IS's project engineering framework.[*] As such, the project began with the creation of a scoping and commitment document that outlined the effort's goals, objectives, and operating assumptions. From the outset, it was agreed that the portal would deliver the following types of content and services:

- Single, integrated, Web-based platform that would serve as an electronic desktop for IS personnel, bringing together in one location all of the documents, information, services, and tools needed to do their work
- Easy access to general IS information, such as news, announcements, events, job postings, and staff profiles
- Easy access to policies, procedures, meeting minutes, presentations, reports, and project-related documents
- Ability to collaborate and participate in threaded discussions online
- Publish-and-subscribe functionality and other automated workflow capabilities
- Web access to Remedy ticket information
- Web automation of IS business processes and forms
- Easy ability to search for and retrieve specific content; standard choices in Web site navigation

---

[*] See *The Hands-On Project Office*, http://www.crcpress.com/e_products/downloads/download.asp?cat_no=AU1991, chpt5~2~project phases~model.

- Secure and reliable repository for document storage and content management
- Ability for users to personalize the KM experience delivered via the portal

All of this was to be delivered through a common, branded IS intranet site that adhered to industry standards and best practices in its Web design, human factors, and workflows. To this end, EO and its external technology partners committed to the creation of a comprehensive portal infrastructure and supporting business and technical processes. The artifacts required as a consequence of these objectives included the following:

- Detailed Web site business and technical requirements, use cases, and architecture documentation
- Schematics of detailed screen workflows, navigation, and page and site functionality
- At least three creative approaches to Web home and secondary page design
- Complete style guide for the chosen Web graphics standard
- Templates for the introduction/depositing of content (Web assets and attachments) and the viewing of content within the site[*]
- Technical documentation for all site workflows and associated applications (applets)
- Business process change documents and process maps to complement anticipated site workflows
- User documentation and training materials

The project team also operated within a framework of guidelines and operating assumptions:

- Application will leverage existing IS technology platforms and adhere to the division's IT architecture standards
- All of the analysis, design, development, and engineered deployment for this project were to be performed by the EO team or outsourced to an external third party
- Most third-party work would be completed off site

---

[*] For the purposes of this chapter, the term *Web assets* refers to any Web pages, frames, content, graphics, or applications that reside within primary or secondary Web pages within a Web site. Other content, such as project artifacts (plans, budgets, scoping statements, and the like), technology white papers, staff biographies, etc., attached to Web pages are summarily referred to as *documents*.

- Depending on the technologies chosen, appropriate IS service units would assist with deployment and subsequently provide tier 2 technical support to the platform solution
- EO would be responsible for initial content delivery and working with other IS content providers
- EO would deliver user and power-user training and would provide tier 1 technical support
- Tier 3 technical support would come from the chosen external partner provider
- Application as delivered will support 200 active, concurrent users within IS (the so-called IS staff experience) with a view toward establishing an IS customer experience at some future date

The choice of an external partner came quickly and serendipitously. Until 2002, Northeastern's systems teams had done little Web development and had not yet standardized on a new E-commerce development platform. A few years before, the university had established Lotus Domino as its primary platform for electronic communication and collaboration. By chance, the director of EO was introduced to a new product: Aptrix from the Australian firm Presence Online.[*] Subsequently acquired by IBM, Aptrix is an application, built from Domino databases, that affords the easy and rapid development of both workflow-enabled Web sites and content management. Because IS had already deployed Domino and had established a center of excellence to support that platform, because there was no competing technology platform already in place, and because the Aptrix product appeared, after careful scrutiny, to offer all the underlying services required at a reasonable price, EO acquired Aptrix in November 2001. Concurrent with that purchase, EO contracted with a Lotus partner, Earley & Associates, to implement Aptrix at NEU and to assist with knowledge portal development.

With Earley & Associates in place, the roles of other project participants became more clearly defined:

- Role of EO — to provide all project management, business analysis and requirements gathering; to be responsible for the maintenance, management, and creation of content for the portal
- Role of IS users — during the development process, to participate in working sessions and focus groups on an as-needed basis; going forward, to be responsible for utilizing and providing content to the portal

---

[*] For more details, see www.aptrix.com or contact info@aptrix.com.

- Role of the IS executive management team — to ensure the proper synchronization among end-user requirements, project plans, and IS delivery; to review and sign off on all project deliverables
- Role of enterprise application technology services (EATS) — to provide technical resources and support concerning the Domino environment and associated services
- Role of enterprise application development services (EADS) — to provide future development services for additional functionality after initial project delivery
- Role of the IS Intranet project steering committee — to review and approve the work of the project team at each key milestone within the project plan

Immediately upon initiating the portal development effort, the project team introduced a three-tiered governance process. First, based on feedback provided by focus groups involving a cross section of IS personnel (and 20 percent of the total staff), the EO team created a project plan, a set of detailed functional requirements, and a series of use cases.* After reviewing this information with Earley & Associates and learning how one might develop these envisioned capabilities through Aptrix, the project team revised its initial PowerPoint portal prototype to reflect more accurately this functionality as afforded by Aptrix. With this refined view in hand, the project team next met with both IS executive management and a representative IS end-user steering committee. The ensuing discussions provided a much clearer picture of requirements for the portal. Up to this point, other than in PowerPoint, not a single employee-hour had been devoted to actual Web development. On the other hand, the team now had a firm understanding of what was needed to deliver a KM service of high value to the IS community.

To document these findings, the PowerPoint prototype itself served as the system of record. The EO team reviewed this document with Earley & Associates in detail. When these reviews proved insufficient, the project team next rewrote the use cases and prepared a more formal set of functional requirements. The latter included the following:

- Glossary of key terms
- Set of general observations and guiding principles concerning site development, look, and feel

---

* Use cases consist of brief scenarios that depict the overall experience and use of Web site functionality. EO's use cases included descriptions of the authentication process, the look and feel of the home page, site navigation, search services, and various Web-enabled business processes (e.g., accessing a team room, applying for a job, accessing IS organization charts, and so forth).

- Key workflows for personalizing the home page, depositing content, revising content, and joining team rooms and discussion databases
- Decomposition of home page functionality
- Description of the universal navigation bar's components
- Decomposition of each Web site service (e.g., looking for a job in IS, subscribing to an IS tool, or updating one's biographical information form) and its associated secondary Web pages

To complement the functional requirements document, the director prepared a so-called Web asset inventory.[*] This table captured technical details concerning the graphic elements and functional components on each page of the envisioned Web site. The inventory included the name of each Web page element (e.g., banner, tag line, navigation bar, text window, drop-down menu), its description, its technical characteristics (e.g., .gif file, URL link, body of text, radio button), and its owner (i.e., the person who would provide that element to the Web site development team).

Taken together, the portal prototype, the use cases, the site functional requirements, and the asset inventory provided the project team with a sufficiently detailed body of information from which to fashion an actual Web site. Next, in January, 2002, the EO team worked in collaboration with Earley & Associates to revise the project plan to reflect the specific deliverables that had emerged from their discovery and analysis efforts.[**] With this input, Earley & Associates finalized its time and cost estimates and then went on to draft technical requirements for the project's Web development environment. The revised prototype was carefully reviewed and approved by both the IS executive management team and the project's steering committee before any further work began. With this approval and with the hardware and software in place, actual construction of the portal could begin.

To make a good start and to avoid scrap and rework down the road, however, it was first necessary for the team to evolve the content and usage frameworks that would ultimately direct the detailed development process. To derive these frameworks, the director of EO and his staff further decomposed work processes for such activities as authoring and submitting documents to the knowledge store, establishing and subse-

---

[*] See *The Hands-On Project Office*, http://www.crcpress.com/e_products/downloads/download.asp?cat_no=AU1991, chpt7~2~Web asset inventory~example.

[**] Aptrix is a relatively new product, especially in the United States. Although it has been adopted recently by Lotus as its Web content management product of choice, Aptrix expertise is in short supply. For Northeastern's IS, the pros of Domino integration outweighed the cons of the team's inexperience with the product. Earley & Associates ended up learning about Aptrix on NEU's dime, but they were careful to bill only for work delivered and not for their applied research.

quently updating staff bios, applying for an IS job opening, and so forth. The team flow-charted each activity and codified its associated rules. EO's KM personnel also constructed a simple database of all IS document types and their associated attributes, including document formats, associated keyword references, and primary authoring and ownership roles.

For each descriptive category (e.g., document title, author, creation date, revision date, review date, keyword description, NEU business unit served, and IS operating unit), the team defined the category and established rules governing its use. Lastly, EO created a holistic view of all IS work,* subdividing tasks into three major groupings:

- Service delivery
- Project delivery
- Back-room operations

The team then further decomposed each task into its constituent parts so that this framework could eventually guide the organization both of the knowledge store and of portal navigation and document retrieval functionality. The first level of that framework as it appears on the new Web site is shown in Exhibit 1.

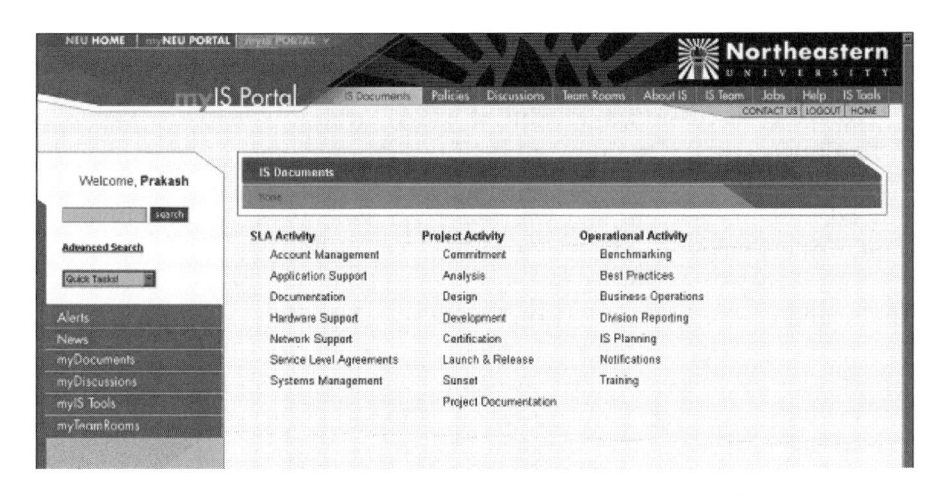

Exhibit 1   Document Categories Organized by IT Function

---

* This view evolved into a detailed taxonomy of terms that added rigor to the site's organization and subsequently served as the basis for document indexing and retrieval. See *The Hands-On Project Office*, http://www.crc-press.com/e_products/downloads/download.asp?cat_no=AU1991, chpt7~3~taxonomy~example.

To be sure, this work constituted the most important set of steps in preparing the team for Web site development. With these frameworks in hand, the team had established the organizing principles for Web site content and the controlled vocabulary, or taxonomy,* for the tagging of individual documents. After the KM team had implemented a series of templates in Aptrix, it became a routine matter to index and store content within the new portal environment. As a starting point, however, the project team first needed to construct the Web site itself.

THE DEVELOPMENT PROCESS: CONSTRUCTING CONTENT AND SERVICE COMPONENTS

In architecting the Web platform for the IS KM system, the project team was very clear that the desired solution must integrate with campus e-mail (Lotus Notes) and offer automated workflows. By choosing Aptrix, the team in effect leveraged the existing Domino platform for e-mail, workflow, collaboration, and multimedia document storage. Within Aptrix, the team implemented its KM taxonomies, document tagging strategies, and Web-enabled business processes. All of these activities contributed to the construction of site components and services but did not yield a Web experience. To that end, the team involved executive management and the steering committee in a discovery process concerning the look and feel of the Web site.

To focus discussion, the team provided steering committee members with a questionnaire to draw out their design preferences and to facilitate their thinking. The team identified several existing Web sites that demonstrated best practices in line with the requirements of the IS knowledge portal. See Exhibit 2.

This survey form yielded the desired results. Within a two-week period, the team achieved consensus on site design criteria:

- Site will adhere to university graphic standards (color, fonts, use of logo)
- Web pages will have consistent NEU and IS branding
- Site will consistently employ a set home page and secondary page format
- There will be no more than three clicks between the home page and the desired secondary page
- Site will include standard and advanced search features and a universal navigation bar

---

* A taxonomy is a uniform, structured system of language applied within a KM process to categorize and catalog content for effective and efficient retrieval by the KM system's user community.

- Site will use a graphics standard with high pixel quality (1024 x 768)
- Site will employ a tab metaphor to integrate look and feel with other NEU Web sites and portals
- Every action item (e.g., notify if changed, download, print), whether regarding a Web asset or a document, must be consistently identified and accessed in the same manner
- In general, there will be no scrolling of content, with some exceptions (e.g., organization charts, full lists of alerts, etc.)
- Design will include screen bread crumbs showing the path from the viewed asset or document to its nested parent location

The project team had enough direction to devise three renderings of potential home and secondary pages. By early February 2002, these representations were ready for steering committee and executive management review for the selection of a final graphic concept for the site. The project's boards of governance quickly agreed on the standards for

Exhibit 2 The Knowledge Portal Design Survey

- Please take a few minutes to browse the following portal-based sites:
  http://www.embedhead.com
  http://www.vci.com
  http://www.bowstreet.com
  http://smallbiz.fleet.com
  http://www22.verizon.com
- While browsing each site, try to put yourself in the mindset of the following types of user and, if possible, provide us with your comments:
  As a first-time user, how hard is it to get started?
  As a familiar user, how hard is it to find something that you know exists, but you do not know the location?
  As an experienced/heavy user, how does this site help get things done?
- In addition to the notes outlined above, please consider the following while browsing the sites, again, providing us with any comments:
  Ease of use
  Clarity of presentation
  Ease of navigation
  Cleanliness of design
- Here are some other considerations to take notice of when browsing and to comment on:
  Words or phrases that capture the IS organization
  Images that should not be considered
  Words or phrases that distinctly do not apply to the IS organization
  Any items that are not covered above
- Of the five sites outlined above, which three did you like the most and why?
- If there are other sites which you feel strongly about, please share them here and provide us with your comments:

186 ■ The Hands-On Project Office

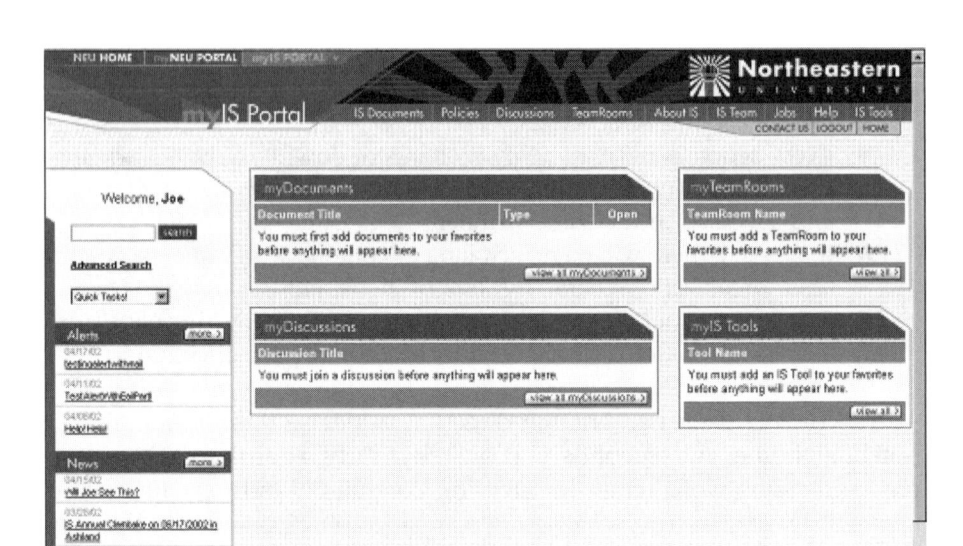

Exhibit 3 The NEU IS Division Knowledge Site

the site's look and feel. With closure on the Web site's style guide, the team now had the direction for formal portal development. This effort began in late March and included the integration of the graphic components, layouts, styles, taxonomies, tables, menus, navigation frameworks, and process workflows through Aptrix. Within six weeks (in early May 2002), the site was ready to receive content. See Exhibit 3.

As Exhibit 3 shows, the IS knowledge portal home page adhered to the standards of design and functionality set forth by the project's governance process. It embraced the university's graphic standards and the preferences of the IS steering committee. The top banner employs NEU's color scheme and university logo, as well as a series of tabs that link the portal (myIS Portal) to the university's public Web site (NEU Home) and Northeastern's forthcoming portal for more generalized university activities (myNEU Portal).

Below the site's banner is the portal's universal navigation bar, with twelve functional tabs, each linking to a key Web site service. This bar appears on the top of the home page and all secondary pages. See Exhibit 4.

Exhibit 4 The Knowledge Portal Navigation Bar

The associated services may be summarized as follows:

- IS documents — provides immediate access to the knowledge store organized according to IS work processes and subprocesses; click on a process task to receive a listing of all Web assets and documents associated with that task; click again to view the document and move it to your home page (myDocuments)
- Policies — offers a complete listing of IS policies and procedures; click to access the policy or procedure and its workflow, to download or print associated electronic forms
- Discussions — provides a comprehensive directory of Domino discussion databases; click to join and add a particular discussion database to your home page (myDiscussions)
- Team rooms — provides a comprehensive directory of Domino TeamRooms; click to join and add a particular TeamRoom to your home page (myTeamRooms)
- About IS — accesses summary descriptions of the IS mission, IS operating units, services, hours of operation
- IS team — accesses an IS phone directory, organization charts, staff search, and updating your own bio
- Jobs — lists all job openings within IS and leads you through a workflow-enabled application process
- Help — provides a site map, frequently asked questions, and a user guide
- IS tools — offers a comprehensive directory of available IS Web-based tools and links to services, such as Notes Web Mail, Remedy, or Lotus Instant Messaging; click to join and add a particular tool to your home page (myIS Tools)
- Contact us — connects you via e-mail to the Webmaster for questions and issues with the portal site
- Logout — logs you off the Web site
- Home — returns you to your personalized home page

The left-hand bar of the home page combines a personalized greeting with the portal's three generalized navigation/search tools and its two push communications services.* See Exhibit 5.

---

* Typically, Web site content delivery is either "push," meaning that the site presents content to the visitor automatically according to predefined rules or audience criteria, or "pull," meaning that the user draws down the information from the site's knowledge store via a search tool or browsing. In the case of the IS knowledge portal, Alerts and News are examples of push services; myDocuments, myDiscussions, and myTeamRooms are examples of pull services.

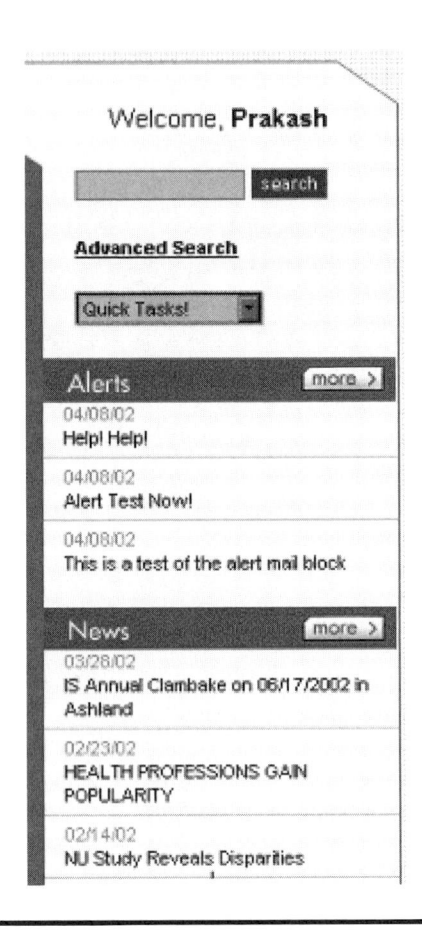

Exhibit 5 Knowledge Portal Search and News Features

In terms of search capabilities, the site offers a standard Web keyword search engine that will, as its outcome, present the user with a list of Web assets and documents relevant to those search terms. Because searching only takes place against taxonomy-based tags and summary document descriptions, each search yields only those assets and documents of highest relevancy to the search. The advanced search function provides actual taxonomy lists from which the user may select search terms, based on document type, document author, IS customer, and IS service provider. Search results may be sorted by document title, author, or date of creation. These search workflows and outcomes are very similar to those employed within Lotus Notes and Domino TeamRooms and therefore were immediately familiar to the portal's user community. Lastly, the site includes a Quick Tasks drop-down list that provides quick and easy navigation links to heavily used Web pages, such as the IS phone directory, the IS organization chart, and the division's list of active business forms.

Below the search tools, one finds Alerts and News. These are the portal's primary information channels where authorized IS personnel publish communiqués of broad interest to the division. In each instance, the most recent entries appear on the side bar, and the More button links the user to a more complete list of alerts or news items, respectively. Alerts concern themselves with system problems and shutdowns, power outages, computer virus attacks, and the like; News typically includes event announcements, fast-breaking news impacting IS, staff changes, and so forth. Together, these two services keep all of IS aware and informed, improving overall team communication and drawing the user community to the site.

The remainder of the Portal home page is customizable by the individual user and serves as his or her electronic desktop, aggregating personal documents with working group resources, services, and tools. All four of the windows operate in a similar manner. For example, when a user finds a document of interest in the knowledge store, he or she has the option to place the document in myDocuments, placing a link to that resource right on the user's home page. Similarly, when reviewing the listings under the Discussions, TeamRooms, and IS Tools tabs, the user may choose to Join (i.e., request access to that TeamRoom, discussion database or IS tool). A request for access is automatically routed by e-mail to the owner of that service, with a copy to the requestor, and a link is added to the appropriate frame (myDiscussions, myTeamRooms, or myIS Tools) on the user's home page. The owner may grant or deny access and inform the requestor accordingly. The link to the service or product will go live when the requestor has received access. At any time, the user may remove a link by deselecting the document, TeamRoom, etc. See Exhibit 6.

Thus, through the IS knowledge portal, IS personnel may organize their work for anywhere, anytime access over the Web. More importantly, the site is the only venue where one may find comprehensive descriptions of IS document libraries, policies and procedures, business forms, productivity tools, collaborative services (i.e., team rooms and discussion databases), and so on. Furthermore, the site's search capabilities facilitate exploration and discovery. Because the site is e-mail–enabled, once the user has identified the author of a document or the moderator of a team room, he or she may seek out that expert for an exchange of tacit knowledge.

The portal includes other services that bind the IS community and facilitate communication and the sharing of expertise. For example, the intranet site provides centralized intelligence on all IS Domino Team-

190 ■ The Hands-On Project Office

myDocuments		
Document Title	Type	Open
Domino Mail Server Outage	Alert	
NU Study Reveals Disparities	News	
How to Add a Document to the IS Documents Library	IS Policy	
NU Employment Application Form	Form	
		view all myDocuments >

myDiscussions
Discussion Title
IS Intranet Portal
view all myDiscussions >

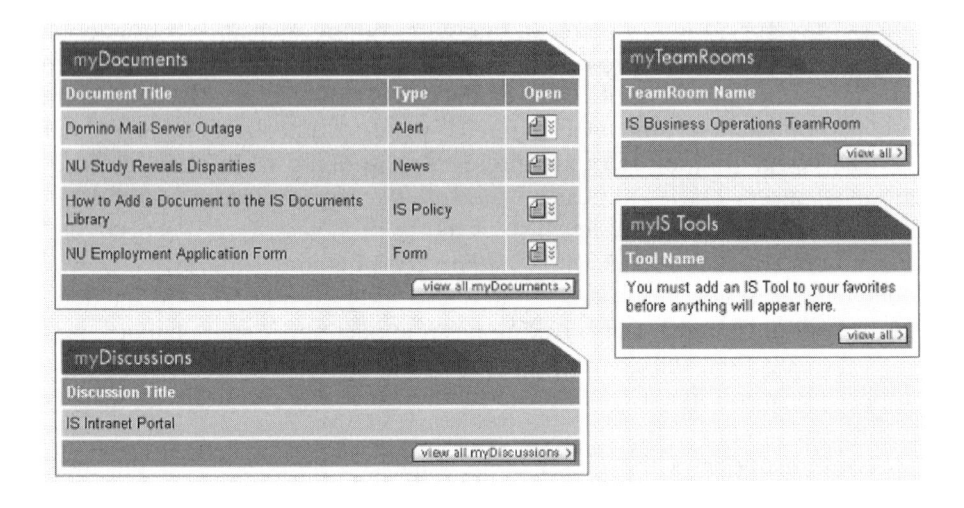

myTeamRooms
TeamRoom Name
IS Business Operations TeamRoom
view all >

myIS Tools
Tool Name
You must add an IS Tool to your favorites before anything will appear here.
view all >

Exhibit 6 The Knowledge Portal's Electronic Desktop

Rooms and discussion databases, making it easy for an interested party to join and to add the link to his or her electronic desktop. Under this umbrella, individual TeamRooms afford an electronic workplace for team collaboration. Typically, IS project and service teams employ team rooms for their respective working documents and supporting information, such as meeting minutes, task lists, research, and analysis. Reference resources, completed documents, and final project artifacts (including final vendor contracts, project plans, and commitment documents), are then promoted to the intranet site at the discretion of the team room's moderator. Each team room is closed to anyone outside that working group; the portal is open to all of IS. Similarly, individual discussion databases offer an electronic forum on a given theme for threaded discussion. The portal assists community members in finding and then joining these discussions.

One of the greatest challenges within an IS organization is knowing where to turn for expert advice. Here, too, the IS intranet provides practical assistance, through a library of individual staff biographies. These bios include basic information, such as the employee's title, reporting relationship, role description, address, and contact numbers. The Web (Aptrix) template provides for an optional digital photo and a curriculum vitae attachment, and allows the employee to select from predetermined lists of skills, skill levels, and job experiences. The data entry process is designed for ease and speed. The resultant tagging ensures accurate retrieval. See Exhibit 7.

Managing Lessons Learned ■ 191

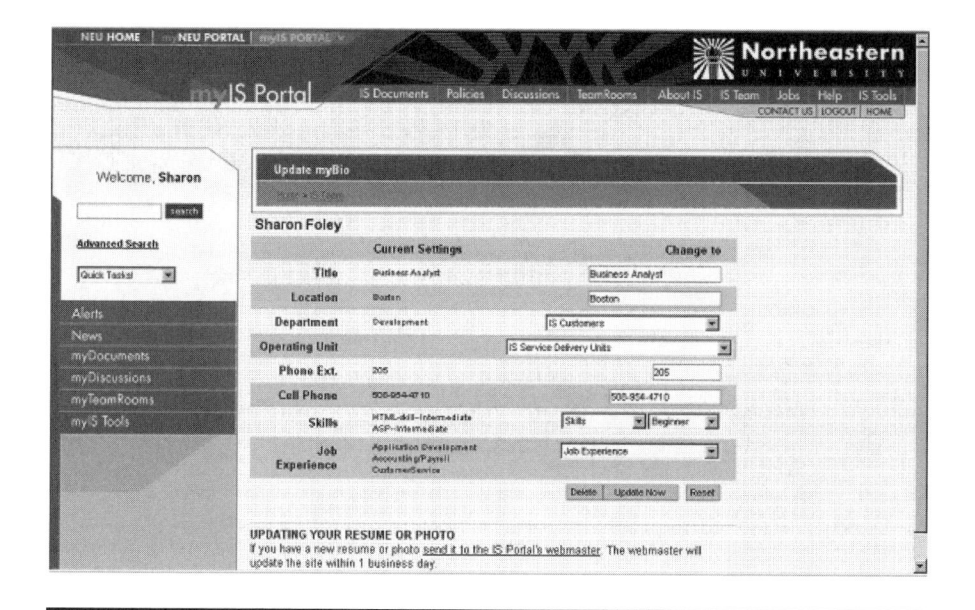

Exhibit 7 The Staff Bio View

Through this portal service, a colleague can quickly discover who in IS is responsible for a particular product or function, or who might have the expertise required for a new project. When you are looking for skills outside the traditional boundaries of someone's job description, this simple tool will serve nicely. Because Aptrix treats individual staff biographies as Web assets, they may be periodically scheduled for review to ensure the currency of the information. Quick Tasks takes the user directly to his or her Bio Update screen, and the site's organization charts are linked directly to staff bios, as well.

Although all of these site features constitute added value, the core benefit of the portal comes from its role as the system of record for IS knowledge. Whether employing the site's search tools or drilling down through the IS Documents tab, the end user's ultimate goal is to obtain relevant and authoritative documents for reference, reuse, or repurposing.* To this end, the site facilitates search and retrieval. Search results are presented as lists. Click on an entry to open the Web container for a document of interest. The container includes the name of the document, the author's name, the document's issue date, the expiration (review) date

---

* For the workflows associated with adding content to the IS knowledge portal, see *The Hands-On Project Office*, http://www.crcpress.com/e_products/downloads/download.asp?cat_no=AU1991, chpt7~4~Document Management~workflows.

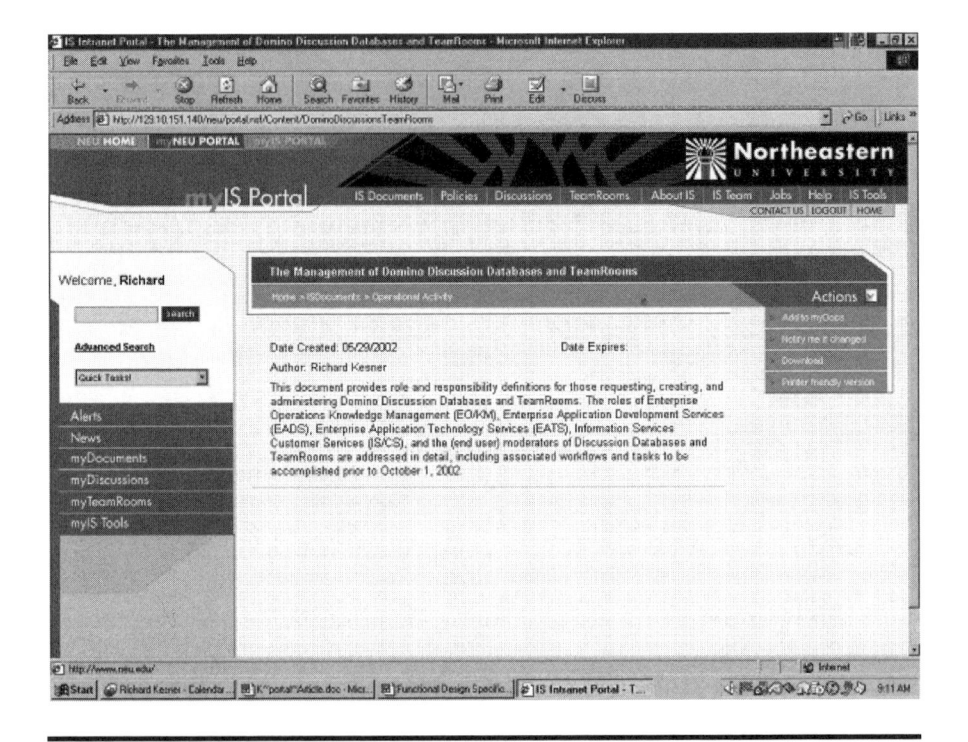

Exhibit 8 The Knowledge Portal Document Container

if appropriate, and a brief description. As a shortcut, the user can simply click on the File Open icon next to the document listing to download the document directly to his or her hard drive.

Once the container is opened, the user will see a standard action bar along the right side of the selected secondary page. See Exhibit 8. This action bar allows the users to take one of four steps:

- Place a link to this Web container and its attachment(s) in myDocuments
- Flag the container so that an e-mail notification is sent to the user if the container or its document is changed by the author
- Download the document to a local or shared hard drive
- Print a version of the Web asset without the header, navigation bar, etc.

The notification function is particularly important for the user who must be informed of changes in a particular IS process, document, or information item. The e-mail–enabled notification includes a link to the appropriate Web container to view those changes. Similarly, if a Web asset or document is scheduled for review (i.e., if it has expired), the site automatically sends an e-mail link to the author to prompt a document review or some other action.

By May 2002, the project team had delivered all of the aforementioned portal functionality and had already relocated much of the IS TeamRoom's content to the new Web site. The team continued to test the validity of the final design through comprehensive site walk-throughs, first among team members and then with the IS executive management team and the project's steering committee. This process took nearly one month, during which time the EO KM team developed more experience in managing the site and in loading a wide range of content. As it added IS TeamRoom documents to the knowledge store, the EO team regularly modified the portal's taxonomy and indexing options. The team also added several utilities to the portal, including WebTrends to track and analyze site usage patterns, and eSurveyor for the automation of end-user data and feedback gathering via the Web and e-mail.

CERTIFICATION, LAUNCH, AND RELEASE

With the development process nearing completion, the IS division's QA process kicked into high gear. Ray Murphy, IS's QA engineer, interviewed members of the portal development team and reviewed the project's documentation to prepare a test strategy and scripts. While this was going on, the project team worked with EATS on the establishment of a test environment. This was no small task, because it required replicating at least part of the IS Domino e-mail directory, IS TeamRooms and discussion databases, and various permissions files. Once this was accomplished, the QA team initiated functional, systems, and performance testing. As testing uncovered problems, these were corrected, and regression testing followed. The QA process took approximately three weeks to complete; in late June 2002, the portal was certified for production.

Concurrently with system testing, the project team prepared for the portal's formal launch and release to the IS division. From the outset of these preparations, the project team and all those associated with the site's development viewed the portal as a potential catalyst for significant change. In their eyes, it would introduce new, untried services, like discussion databases, and a staff bio library. It would impose added rigor to intradivisional communications and document management. And it would call upon each member of the division to operate in a more integrated fashion as part of the larger IS community. Clearly, not everyone in IS had enrolled themselves in all these changes, nor was the PMO entirely prepared for the added work and responsibilities that would fall to that team. The launch effort would need to address the many different dimensions of these issues if the portal and its underlying KM processes were to be embraced by the greater IS team.

To begin, the project team initiated a marketing campaign. Members of the steering committee, who represented a cross section of IS, were asked at each milestone to talk up the project with their colleagues. As the site moved into production, the director of EO, accompanied by a panel drawn from the steering committee, provided a brief overview of the portal and its benefits to the entire division. Terms like *knowledge management* and *business process reengineering* did not enter into this presentation. Instead, the script addressed how the portal would add value to IS personnel in their day-to-day work. As a follow-up to these large sessions, the EO KM team met face-to-face with individual IS operating units for a more detailed tour and a more focused discussion of how the site might address the particular needs of that working group. Mouse pads bearing the portal home page and URL were distributed at these sessions. Additional marketing programs were planned throughout the first year of the portal's operation to promote use and the submission of content to the site.

Although the portal was designed to be intuitive to the end user, IS staff training and documentation were still important. EO viewed the departmental meetings as the project team's opportunity to deliver end-user training. The site itself includes a Web user's guide and a list of questions frequently asked of the Webmaster. Even so, the team anticipated that most users would learn about the site by exploring its functionality. To address individual problems and questions, the portal provides both e-mail and instant messenger links to the Webmaster. By contrast, power users (i.e., those users who are major content providers to the site) received formal training in Aptrix, as well as appropriate supporting documentation. Fortunately, given the choice of Aptrix, the platform looks and operates very much like other Domino services, such as e-mail and team rooms, making the training of power users that much easier.

ONGOING OPERATIONS

By the summer of 2002, after three months of design and analysis, two months of development, and one month of QA/RM, the IS knowledge portal was launched. The marketing to and training of users ran throughout the summer and into the fall, the start of a new academic year. The knowledge store was fairly substantial at launch but grows considerably each quarter as both EO and a cadre of IS power users regularly adds content to the site.

The ongoing support and maintenance of the IS knowledge portal is now shared among various IS division operating units:[*]

- Enterprise operations KM (i.e., the Webmaster) — provides tier 1 support for all portal users and serves as the liaison between most content providers and the knowledge store. EO also maintains the overall rigor of site taxonomies and services and implements minor enhancements to the portal's offerings, the strategic direction of the service, and its ongoing promotion and proliferation.
- Enterprise application technology services — maintains the underlying Domino platform, installs Aptrix patches and upgrades, and otherwise provides tier 2 support for the application.
- Enterprise application development services or Earley & Associates — provide tier 3 support and any major additions to current portal offerings.
- Power users — enter, update, tag, index their own content within Aptrix.
- Other content providers — direct their new or updated Web assets and documents to the Webmaster for posting to the knowledge store.
- Team room, discussion database, and IS tool moderators — detemine access rights to their respective information resources and services, which they ensure remain current.
- Many within IS — employ the new KM environment and electronic desktop in their daily work.

Beyond ensuring that the portal platform remains robust and that the experience continues to meet the needs of the IS user community, the IS PMO works with all of IS to make certain that the IS intranet site remains fresh, current, and comprehensive in terms of its content and services. The PMO team recognizes the need for ongoing marketing and continuous improvement of the service and hopes that its colleagues will take up the ongoing challenge of leveraging this new KM service.

---

[*] Current conversations point toward the reengineering of myIS in terms of both its Web experience and its underlying technology. The emerging plan calls for moving the knowledge portal in a redesigned form onto the University's open standards (uPortal) Web portal (i.e., myNEU), employing SCT/Campus Pipeline's Luminis Portal and Documentum Content Management Server technologies.

LESSONS LEARNED

It is too soon to tell whether the NEU IS division's venture into Web-enabled KM has met the expectations of its creators or their customers. The end product is not static and will evolve with use over time. Nevertheless, much has emerged from the early stages of this effort. The IS team can point to the following noteworthy outcomes:

- Single, integrated, Web-based platform delivering anywhere, anytime access to authoritative IS operational information, project documentation, business forms and process, and best practices
- Better control of IS documents, standards, and best practices
- Improved staff communications and electronic venues for the open discussion of key issues impacting IS
- Improved collaboration across project and service delivery teams
- Content-management system that encourages regular, systematic review and revision of IS business practices and technical standards
- Test case/laboratory for KM and portal functionality

This case study affords a number of relevant lessons learned. First of all, the same forces that drove Northeastern University to establish a PMO justified the investment in the creation of a knowledge portal. Just as the PMO serves the IS division as a center of expertise in service and project delivery, the portal provides an access point for the tools, templates, and process artifacts supporting IS team efforts. The PMO team acts as the archivists, anthropologists, and guardians of the division's tacit knowledge and its collected best practices. The portal captures the more explicit evidence of what works and what does not. Second, perhaps key to the PMO and the portal, both perform aspects of a clearinghouse function for the activities of the greater IS organization. The PMO team keeps an eye all on aspects of service and project delivery. It also looks for ways to rationalize and streamline new assignments through the leverage and reuse of past work. For its part, the portal archives service and project artifacts so that these may be adapted and redeployed by the PMO and its customers as these knowledge assets apply to similar tasks.

Thus, the value propositions for the PMO and the knowledge portal are intertwined. Through its efforts, the PMO creates and manages the processes and documents that would feed any knowledge repository. It has both the perspective and the appropriate analytical disposition to create a knowledge store and to maintain it thereafter. Unlike the vast majority of their IS organization colleagues, PMO staff members enjoy enough separation from day-to-day delivery to serve objectively as the keepers of the unit's knowledge, choosing what to save and how to repurpose it for reuse. At the same time, PMO personnel are intimate enough with the details of IT organization work to identify documents of value, properly index these for retrieval, and establish complementary KM services for instances of unrecorded (i.e., tacit) learning.

To achieve these objectives and to benefit the entire IS organization, the knowledge base places special demands on both the IS management team and the PMO. Even though the myIS portal was shaped by the direct requirements of Northeastern's IS rank and file, as of the date of writing, widespread adoption has not occurred. Like their line-of-business colleagues, the university's technologists are slow to adopt new work practices that leverage the knowledge site's capabilities. Staff members continue to rely on content stored on their own computers rather than to refer to the common information library on the knowledge portal, and they find other, less economical ways to achieve the outcomes enabled by myIS services. By contrast, the PMO (and to a lesser extent IS management) regularly employs the portal in its work. In the final analysis, NEU's first KM application may be rated as a successful prototype to what is sure to be a successor offering that will emerge from the university's Web services platform.

Concerning the outcomes of this case study, the author's perspective is hardly free of prejudice. Nevertheless, I offer my measure of the tangible advantages to an IT organization in developing its own knowledge store:

- As a resource library for standardized forms, workflows, and checklists, an IT knowledge store promotes consistency of practice and perhaps more predictable results in service and project delivery.
- As a centralized store, the portal saves your team time and effort in the search for reusable knowledge artifacts. It may even avoid the regular reinvention of past practices (and mistakes).
- As an archive of best practices, success stories, and proven tools, the store serves as a training platform for new hires, technical personnel transitioning to management, and service delivery and project teams searching for what works.
- This same store serves as the toolbox for your PMO team's project managers and business analysts.
- In the face of increasing scrutiny by your colleagues on the business side of the house, the application of knowledge store assets will demonstrate your commitment to greater efficiency and effectiveness in service and product delivery, to continuous improvement, and to doing more with less.

Of course, to achieve these lofty outcomes, some group, like the PMO team, must maintain your IT knowledge store and continue to enhance its scope of services and breadth of content. This is an evolutionary process. If responsibilities are distributed across the organization, the establishment of a KM function need not be resource-intensive. However, maintenance of a KM function does require real discipline and attention to detail. Furthermore, if it is to succeed, the portal requires the ongoing and active support of IT management. By support, I mean the regular, public use of and advocacy for the knowledge store as a key element in the way the IT organization does its business. To my mind, your investment will pay a healthy return.

8

ARCHITECTING SUCCESS — THE ROLE OF SENSIBLE IT ARCHITECTURE MANAGEMENT IN SUCCESSFUL SERVICE DELIVERY: A CASE STUDY

INTRODUCTION

Chapter 7 examined the role that knowledge management (KM) can play in enabling IT organization performance. As that case study illustrates, PMO personnel can help leverage the assets of a IT knowledge base to foster collaboration and communication within the enterprise's IT organization. Through the repurposing and reuse of knowledge artifacts, service and project teams will save time, avoid past mistakes, and build on past learning. But KM may serve other ends than IT process improvement. IT professionals also need help in working with vast bodies of technical information about the computer hardware and software deployed across the enterprise. Here, too, KM can make a difference by supplying a framework for the creation, organization, maintenance, and dissemination of data about the enterprise's IT products, whether planned or in play.

Chapter 8 explores a KM approach to IT architecture planning and management. The PMO team is the body of IT professionals best positioned to gather the information and associated artifacts to support the envisioned architecture process. And, like Chapter 7's case study, the example that follows illustrates how PMO personnel, employing a Web infrastructure, can engage technical experts from across IT to draw out this knowledge. In this chapter's example, a PMO-facilitated team leverages the explicit and tacit technical knowledge of New England Financial (NEF) personnel to create an accurate depiction of the current state and, more importantly, to extrapolate from this foundation, deriving a vision and plans for the extension of the enterprise's IT architecture.

The rapid pace of business change and IT innovation demands highly flexible and responsive, yet strategic, planning and procurement processes. For that matter, IT's ability to deliver services and projects successfully depends, to a certain extent, on the availability of the right technology architecture. Thus, the theme of IT architecture planning integrates well with the themes of delivery management and PMO enablement. Unfortunately, many IT organizations, burdened with the responsibility of meeting the day-to-day requirements of the enterprise, have all they can do simply to keep up with immediate needs. As a consequence, technology planning and investment strategies tend to be reactive and focused on a short-horizoned return on investment (ROI). In these circumstances, more long-term strategic thinking and research gets set aside as the enterprise's IT leadership deals with the pressing challenges of the moment.

Yet, year after year the IT industry trade press reports that the paramount concerns of CIOs are:

- Aligning their IT investments with their organizations' business strategies
- Containing technology costs while leveraging IT-enabled process improvements

How does one balance this need for immediate IT-enabled results that fall to the corporate bottom line with the corresponding need to position the enterprise's technology infrastructure to anticipate future business requirements? And how can this work get done without adding to staffing costs or detracting from immediate IT team commitments? Concerns for the present management and for the future planning of an IT architecture need not be mutually exclusive. Indeed, the NEF case study will describe how an organization's IT architecture process fueled the continuous updating and improvement of its technology investments, even as it positioned the enterprise's future technological capabilities. Furthermore, this was achieved without adding personnel and while engaging a cross section of the IT organization in technology planning.

This example embraces a practitioner's approach as tested and proved in a real-world work environment.* Throughout, I focus on the processes required for effective, efficient, and comprehensive IT architecture planning aligned with the business objectives of the enterprise. As noted previously in this volume, rather than endorsing rigid application of my framework, I encourage the reader to reflect on the following narrative and adapt its examples and suggestions to the particulars of his or her work setting. In the final analysis, the success of IT architecture planning relies on the practitioner's clarity of vision, intellectual rigor, and commitment to a sustained process. Excellence in these activities will inevitably lead to the judicious use of enterprise resources in the identification and procurement of appropriate information technologies.

FRAMING THE IT ARCHITECTURE PLANNING AND MANAGEMENT PROCESS

N. Dean Meyer, a highly regarded IT organizational consultant, has observed that "the very purpose of architecture is to allow opportunistic, business-driven implementation efforts while still evolving toward an integrated product line. ... Architecture is a set of standards, guidelines, and statements of direction that constrain the design of solutions for the purpose of eventual integration."**

In these few words, Meyer captures the essence of what the IT architecture process is all about and indicates its worth to the enterprise. For Meyer, an IT architecture is not an archive of past and present IT purchases; rather, it is an integrative process that influences purchasing decisions while driving the enterprise to a higher level of IT integration and enabled performance excellence. His recognition of the centrality of the architect's role is now being mirrored in the hiring practices of large and small IT organizations worldwide. I met Meyer early in my NEF years. His writing influenced my own thinking about the high-level process that came to define NEF's IT architecture management practices.

According to Meyer, at its core an IT architecture provides four primary sets of services. First, it offers a comprehensive, systematic consideration of the enterprise's current IT investments. Given that most

---

* For a period of time, the author served as the vice president of planning of New England Financial Information Services (NEF/IS), the IT arm of New England Financial, a member of the MetLife family of companies. Among his responsibilities, the author was responsible for business and technology planning, as well as the IT architecture process for NEF/IS. He was assisted by Tor Stenwall, senior technical architect for NEF/IS, now retired.

** N. Dean Meyer, *Structural Cybernetics* (Ridgefield, CT: N. Dean Meyer and Associates, 1995), 59-60, 65.

IT organizations are notoriously lax when it comes to documenting their work, this is no mean task, involving the collection of detailed information about existing legacy systems, *de facto* corporate IT standards, and the history of IT systems and operations. Second, the IT architecture process provides a framework for the planning, evaluation, and coordination of an enterprise's future IT investments in that it articulates current and anticipated technology standards, as well as the overall direction of IT procurement. Indeed, if properly executed, the architecture will offer a clear and cogent picture of the enterprise's technology future. The ultimate objective of these efforts is the seamless integration of information technologies in support of the business.

Third, the architecture acts as a filter for the assessment of IT options. As new technologies emerge, the enterprise's IT leadership must be in a position to determine whether such events afford opportunities. Given the pace of change and innovation within the IT industry, this effort alone could consume all of IT management's energies. A comprehensive architecture model will assist the team in its triage of IT product and service announcements and in plotting a sensible course through myriad offerings. It also acts as a means to discipline applied research and otherwise whimsical exploratory forays into emerging information technologies.

Lastly, the IT architecture process establishes a platform for the identification of IT innovations that marry well with enterprise business requirements or that create whole new business opportunities. Here, the architecture process's contribution is highly proactive, allowing IT leaders to advocate for particular technology investments and business process changes. Thus, an IT architecture chronicles and manages the current state of the enterprise's technology and serves as the basis for the planning and integration of future initiatives. It provides a context and a framework for assessing the gyrations and offerings of the IT marketplace and even, in time, a platform for advocating change. To demonstrate these points, the author will now turn to a case study for the development of an IT architecture process at New England Financial. Although begun at NEF, this process has been adopted and adapted by NEF's parent company as one of the ways in which MetLife now manages its IT infrastructure.

INTRODUCING THE CASE STUDY

New England Financial provides comprehensive insurance and investment products to about 2.5 million customers through a national field force of dedicated and independent agents and financial advisors. Headquartered in Boston, Massachusetts, and numbering several thousand employees, NEF is one of MetLife's flagship product lines. Like its parent company,

NEF supports a large number of complex, proprietary information systems that enable its diverse product offerings. Primarily a user of mainframe-based information technologies, NEF has moved with the industry over the last decade into client-server and Web-enabled systems for customer servicing and product management. As a result, almost any flavor of desktop, middle-tier, or back-end technology may be found within NEF's data centers, home office, or branch agencies.

For NEF's IT organization (NEF/IS), managing the complexity and diversity associated with this embedded base of IT is a formidable task. Tracking the location and versions of hardware and software presents its own special challenges, as does finding the owners and technology thought leaders associated with particular IT products or services. When the author joined NEF/IS in 1997, there was no master plan for the migration of existing IT or the introduction of emerging technologies. Furthermore, there was no systematic process for the assessment of IT options and the creation of a long-term IT strategy. Last but not least, there was no road map to identify internal experts with the systems over which they held responsibility. Finding out where to turn required multiple phone calls, e-mail messages, and a lot of unproductive time.

This is not to suggest that NEF/IS was poorly managed. It merely suffered the fate of any large IT organization growing in response to the various internal and external forces driving the parent enterprise's IT investments. In fact, the chief executive officer at the time, Robert Shafto, had formerly served as the first CIO of NEF, and his successor as CIO, Greg Ross (now retired), was one of the finest IT executives under whom I have served. These enterprise business leaders had a clear sense of what IT could do for them and had developed a cogent picture of IT's future within NEF. For its part, the IT organization was very much aligned with the business side of the house. Generally speaking, NEF/IS also enjoyed a solid record in terms of IT service and project delivery. Nevertheless, high-level architectural discussions were not informed by a detailed picture of the current state of NEF/IS investments, nor did the rank and file of the IT organization know where to turn for expertise on particular technologies. With the MetLife merger, it became increasingly important to coordinate and standardize IT purchases. As part of my role within NEF/IS, I was asked to address this gap in the IT organization's planning capabilities and corporate knowledge.

THE UNDERLYING ASSUMPTIONS OF AN ENTERPRISE'S IT STRATEGY AND ARCHITECTURE

In approaching this assignment, I sought participation and input from across the NEF/IS organization. My objective was to engage as many of the firm's IT thought leaders as I could find and to employ them as my board of advisors in crafting a model for the definition and depiction of NEF's IT architecture.* My first task in working with this large and diverse group of contributors was to draw from them a set of framing principles for the overall architecture authoring process. The specific nature of these assumptions may vary, of course, according to the industry and technology contexts of the enterprise. Nevertheless, the general character of these assumptions will have much in common from organization to organization. First and foremost, they must be clear, relevant to the enterprise, and actionable. Second, they must reflect the values and strategic direction of the enterprise's IT organization.** Third, they must speak to all stakeholders of the process, including the enterprise's business leaders, its IT team, and IT's external technology partner providers.

In general, the architecture process's underlying assumptions should address three areas of concern:

- Articulation of the relevant business/industry context and perhaps of high-level business strategies
- End user–focused principles
- Technology-focused principles

The following illustration from the application of this approach at NEF/IS offers a set of process assumptions that worked for that particular context but may be tailored to better fit the needs of the reader's organization. Collectively, these assumptions serve as a guide for the players in the planning process and as the framework within which IT architectural choices may be described, categorized, filtered, and ultimately selected. See Exhibit 1.

In this example, the first set of assumptions speaks to the necessity of mapping IT choices to the requirements of the business. Although this may seem obvious, the statements that follow elaborate on the particulars, such as the need for information technologies that foster collaboration among all players along the enterprise's value chain of delivery. The example next includes language concerning the management of financial data and reporting (of particular importance to a business like NEF), the protection of the enterprise's data assets, and so forth.

---

* This process transferred to New York City once MetLife assumed control of NEF's IT direction.
** Some CIOs choose to work with their management teams and business partners to frame architecture process assumptions, while others choose to delegate this task to the architecture team itself. In either case, the enterprise's leadership ultimately must endorse the principles upon which the architecture is built to legitimize the process. PMO personnel might serve to collect and codify this information for executive and architectural planning process review.

Exhibit 1   IT Architecture Process Assumptions

BUSINESS PRIORITY ASSUMPTIONS:

- IT architecture decisions will take into account the business practices and standards of the enterprise and those of its industry competitors.
- IT investments will align with the enterprise's business strategies.
- IT will facilitate the conduct of business and the retention of valued (internal) employees and (external) business partners.
- Easy access to financial information required in running a public company is a paramount concern.
- Data quality and integrity are essential to the success of the enterprise.
- Customer and business partner self-servicing are key business strategies.

USER-FOCUSED PRINCIPLES:

- IT's primary focus is customer-driven product and service delivery, with an emphasis on customer value and time to market.
- Responsibility for the implementation of IT, as well as the security and integrity of data, is shared among the users of enterprise IT and the corporate technology team.
- Electronic information in all formats is a vital asset of the enterprise and must be valued, protected, and leveraged.
- A primary focus of IT product and service delivery is to enable and empower users via self-directed, anywhere, anytime access to electronic information assets and services.
- No redundant data entry or storage should be allowed, except as dictated by technological or business requirements.

TECHNOLOGY-FOCUSED PRINCIPLES:

- The IT organization should purchase and integrate, rather than build from scratch, IT products for generic functions.
- IT should build applications for strategic advantage when dictated by unique business requirements or opportunities.
- IT should favor individual best-of-breed products over integrated product suites.
- IT should avoid customization of generic products except through supported interfaces (APIs).
- IT should support a common, homogeneous technology platform while recognizing the need for renewal and replacement.
- IT should sunset older, marginalized technologies as soon as possible.
- IT should migrate to an open, standards-based architecture, using Web-based, thin-client, server-centric applications that enable ease of application distribution, maintenance, and support.
- IT should ensure secure and confidential transaction processing and information sharing.

The second grouping of assumptions speaks to the end user's experience. Here, the focus is on those values that enable the independent action of the enterprise's employees in serving their customers. This section also emphasizes the need for systems integration, so that data entered once applies to all systems that require it. Together, these statements map an end-user IT experience and may be employed as a set of validation rules when interacting with your business partners to select among IT options. The final set of assumptions addresses technical considerations in the selection of technologies, including the preference to buy rather than build systems and to choose best-in-class solutions over system suites. Even these few sentences draw clear boundaries around acceptable IT choices.

The example includes powerful statements that might shape any organization's selection of information technologies. Clearly, not all of the particulars in Exhibit 1 will apply to the reader's work environment, and perhaps these statements ignore issues that may be of utmost concern. On the other hand, the NEF assumption list contains the key components of a framework for building a tailored set of assumptions for your enterprise. In the first instance, they focus upon those business issues of paramount importance to the business. Although the industry benchmarking of IT deployments may be a temporary fixation on the part of your bosses in management, meeting the core needs of the business should always be the primary driver of IT investment and hence of the corporation's IT architecture. The opening statements in Exhibit 1 make this case clearly.

Second, the assumptions speak to what end users can expect from enterprise IT and what is expected of them in return. Building a culture around the effective use of IT is essential to the success of any technology deployment. If you ignore the values and norms of your enterprise users, you may find yourself investing in technologies that are underutilized by your organization's work force. For example, your enterprise may standardize on a Microsoft Windows platform for the desktop, but it must also recognize that some business units, like publishing and graphic design, may be better served by an alternative product. Lastly, in addressing key technology strategies, such as buy versus build and a focus on open standards, your assumptions will help to frame the playing field. Within NEF/IS, these value statements framed the process, but they required more detailed elaboration as they were applied to particular categories of IT products and services. As the reader will observe, my process addressed these needs in due course.

By involving your organization's PMO personnel, CREs, IT team technical experts, and IT management in these deliberations, you can explore and ultimately identify those IT architecture assumptions that are right for your business. When conducting this work for NEF, I involved a representative cross section of IT thought leaders and, in a few instances, so-called power users. I kept management out of the more detailed process deliberations to allow my colleagues on the IT architecture advisory board free rein. When we reached agreement, representatives from the team brought our proposals before IT executive management for their review and approval. This process did take time, but it was an essential investment. With a firm consensus around these statements, the IT architectural planning team had an effective filter to employ when considering choices and weeding out options. The foundation was now laid for the process of articulating an IT architecture for NEF.

THE PROCESS OF BUILDING AND MAINTAINING AN IT ARCHITECTURE

An architectural process should be highly collaborative, involving a cross section of IT management and technical experts. To initiate this process, draft a team charter that clearly defines roles, responsibilities, and process deliverables. The elements to be addressed in such a charter may be summarized as follows:

- Process sponsor — typically, the enterprise's CIO and his or her executive team. Process sponsors must sign off and endorse the process. Without their clout, you may not get buy-in and hence the time and attention you need from rank-and-file IT thought leaders.
- Process customers — the enterprise's IT team. By its very nature, the IT architecture is highly technical and not meant for the use of the IT organization's customers (i.e., end-users). Rather, it is employed by the IT team proactively to meet the needs of its customers.
- Process participants — an IT executive leader (as project director), one or more IT architects (as technical and integration experts),* members of the PMO (as process facilitators and knowledge managers), and various technology domain teams (as content providers).**
- Process deliverables — the creation of a clearly articulated and comprehensive framework for all the technology currently deployed

---

* In smaller IT organizations, there may not be a designated individual in a formal architect's role. On the other hand, you will find that others perform some aspects of this function as part of their current infrastructure, systems development, or data management assignments. Involve these people in the process.
** As the reader will observe, the author has organized his IT architecture efforts around a series of technology domains (e.g., Web and Internet technologies, server and storage technologies, and networking technologies) that draw on the various centers of IT specialization across the enterprise featured in the case study.

within the enterprise, indicating what is in place, sunsetting plans, future directions, established and anticipated standards, and so forth. This outcome will be enhanced if it also identifies those persons responsible for the IT systems, services, and products mentioned in the architecture documents, as well as their contact information. Furthermore, it is essential that the process embrace an outcome that sustains the IT architecture process itself over time.

- Process timetable — two phases. Phase 1 (6–12 months) involves the initial establishment of the architecture; Phase 2 involves ongoing, continuous maintenance of the framework. Your timetable will depend on the urgency of your organization's need for an architecture process and the availability of resources you can devote to this effort. In the case of NEF, process participants worked this project into their already-packed schedules, hence its elongated delivery timeframe.
- Resources required — a limited amount of dedicated, high-level technical expertise through the assigned IT executive and his or her architect(s); ongoing support from the IT organization's technical teams as part of their business-as-usual task work.*

As indicated, the process must come with the full support of the IT organization's management team. Without its good will, the architects will have no access to the attention and expertise of those within IT operating units whose input is necessary for the development of architecture knowledge components. The success of the overall process is contingent upon drawing out knowledge from a significant cross section of the IT organization and in keeping those people involved in the ongoing maintenance of process deliverables. Because the ultimate audience for architecture documents will be the IT organization itself, the process must communicate effectively and economically to this audience. Thus, involving the technology thought leaders within your organization — and keeping them involved — is essential to the overall quality of the end result.

As an organizational principle, most architecture processes are built around various technology domains or areas of specialization. These groupings allow process participants to focus on and build a critical mass of knowledge around those categories of IT products and services that matter to the enterprise. Those who consult in the area of IT architecture planning have strong feelings about the number and description of these domains. The author does not

---

* Bear in mind that this sort of work is highly rewarding to technologists, especially when they see the fruits of their efforts realized in the statement of enterprise IT standards. Most gladly devote the extra time required, as part of their overall development as professionals and for the privilege of being treated as thought leaders within the organization.

Exhibit 2 An IT Architecture Process Technology Domain Design

- Application development — encompassing application development and testing tools, methodologies, templates, frameworks, application technical design, and an application development infrastructure for messaging, component reuse, etc.
- Business applications — encompassing the total life cycle of the portfolios of manufacturing, distribution, and corporate information systems employed by the enterprise
- Data — encompassing database design tools and methods, directory technologies, database management systems, and data administration tools and systems.
- Desktop and collaboration — encompassing desktop, laptop, and personal data assistant (PDA) hardware, operating systems, and productivity software; also e-mail, calendaring, groupware, and related technologies
- Human factors — encompassing design methodologies, graphical user interface guidelines and standards, usability testing techniques and tools, and Web authoring tools and design standards
- Knowledge management — encompassing data warehouse and mining, online analytical processing (OLAP) tools and methodologies, and knowledge capture, storage, and access
- Network infrastructure — encompassing telecommunications and local and wide area networking technologies
- Security — encompassing IT security technologies
- Server/mainframe — encompassing departmental and enterprise-level servers, storage systems, printers, scanners, backup systems, etc.
- Voice — encompassing telephone switches, voice mail, voice response systems, call-center technologies, etc.
- Web and Internet services — encompassing Web-specific products and services that either stand alone or enable other enterprise IT products and services

share this concern. In brief, the organization of an enterprise's architecture process must be driven by its own business and technology needs. As long as the structure encompasses all of the technologies in play within the organization, it suffices. For these reasons, within NEF/IS the author employed an IT architecture advisory board to derive an IT domain model for NEF. Through a series of brainstorming exercises, the process team came up with a list of IT architecture domains. The reader's list may differ. The litmus test for its appropriateness is that when you are done, the list must embrace all those technologies that reside within your business. See Exhibit 2.

Again, this list is merely illustrative, serving as an organizing framework for those laboring within the process. Each domain will have its own team leader, preferably the most senior in-house expert in that particular subject matter, and should include a healthy cross section of domain-specific representatives. In the case of NEF, the author, as the executive responsible for

210 ■ The Hands-On Project Office

Application Development Application development and testing tools, methodologies, templates, frameworks, application technical design, application infrastructure for messaging, brokering, reuse, TPM's.	**Knowledge Management** Data warehousing, data mining, OLAP tools and methodologies. Knowledge capture, storage, and access/sharing tools, methods, guidelines and standards. C.K.O. recommendation.
Business Applications Manufacturing, distribution, and corporate applications: sunsetting, componentization, application infrastructure exploitation (messaging, reuse, templates, etc.)	**Network Infrastructure** Telecom. hardware, software, usage policies, procedures, and systems management (alarms/ problems/ changes/ capacity/ performance/ configuration/ security/ disaster recovery).
Data Architecture Database design tools and methods, directory technology. DBMS's: legacy, distributed, desktop, replication. NEF business data complexes: administration and evolution.	**Security Architecture** Policies and procedures for identification, access, accountability, confidentiality/ privacy, and data integrity. Security awareness programs, security administration procedures and publications.
Desktop & Collaboration Desktop hardware & operating systems, laptops, PDA's, browsers, Office Suites, web authoring, E-mail, calendaring, discussion groups, teleconferencing, e-forms, workflow.	**Server/Mainframe Infrastructure** Departmental and Enterprise level hardware/software (servers, DASD, tape, printers, scanners, CD-ROM, backup, O/S's), output management, and systems management.
Human Factors Design methodologies, GUI guidelines and standards, usability testing techniques, Web design guidelines and standards, Web authoring procedures, high-end web authoring tools.	**Voice Technologies** Voice mail, voice response units (VRU's), voice recognition, PBX's, computer-telephony integration (CTI), connectivity: local, long-distance, 800-number, tie lines, ACD's, Web telephony.

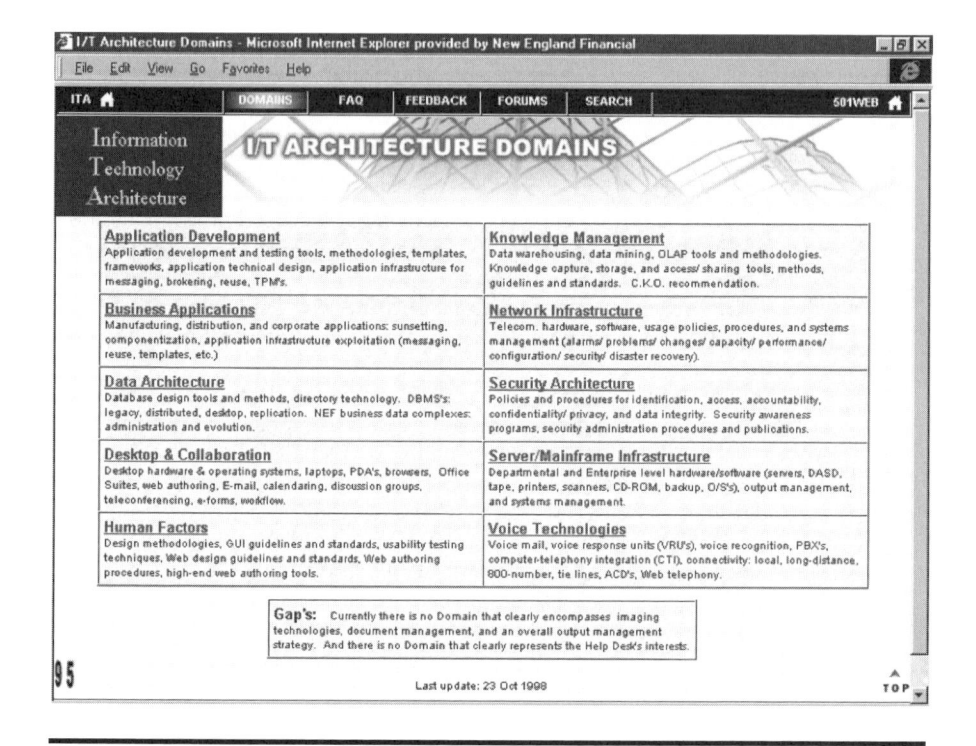

Gap's: Currently there is no Domain that clearly encompasses imaging technologies, document management, and an overall output management strategy. And there is no Domain that clearly represents the Help Desk's interests.

Last update: 23 Oct 1998

Exhibit 3 The NEF/IS Architecture Domain Teams

the overall architecture process, served *ex officio* on each domain working group, as did the IT architect. Together, we acted as process guardians, facilitators, scribes, and timekeepers of domain team deliberations. We also worked as the integrators of domain team activities and the creators of the artifacts and KM systems emerging from their collective efforts. In the reader's case, PMO personnel may serve alongside or in the place of IT architects to keep these efforts on track and to chronicle results. See Exhibit 3.

Although product and service managers may participate on teams (as they did in the case of NEF), it is recommended that the working groups primarily include those with direct, hands-on experience with the technologies in question, as well as those who have a strong interest in the evolving body of knowledge encompassed by each product or service domain. Some organizations will separate the operational servicing of a technology from related applied research. The domain team should include representatives from both functional areas. At the same time, no domain team should exceed ten members. Otherwise, process overhead will become burdensome, slowing team decision-making. Bear in mind that, in other ways, the architecture process is inclusive. Any number of additional technologists may be solicited for their input without requiring them to serve in a working group. Most importantly, each domain team is an advisory/advocacy body to focus data collection and conversation. The architect (or the PMO) serves as the auditor of enterprise compliance with architectural standards, while the corporation's IT operations managers serve as implementers of the actual technologies once those are chosen.

After the executive NEF/IS leadership reviewed and approved the process charter and the organization and membership of the architecture domain teams, the author and his colleagues received a mandate to proceed. From the outset, I made it clear to those participating that this process should be integrated with their day-to-day work. Our underlying work assumption was that as they used technologies in-house and as they followed developments in their fields of expertise, they would also apply this knowledge to building and documenting an IT architecture framework for NEF. Thus, it was our hope and expectation that the process would merely formalize what these IT specialists had been doing all along. To facilitate these outcomes, the author and his colleagues, who constituted a virtual PMO team, provided process management, simple information-gathering tools, and lots of encouragement.

Freed from the detailed administrative and KM minutiae that naturally accompany an IT architectural process, the domain teams focused on grass-roots information gathering, communication, and data analysis concerning the IT products and services in play within NEF. With the aid of the process tools provided by my office, the teams grew a body of knowledge that addressed the following areas of interest:

- Domain principles — these include any particular principles and assumptions that shape the architectural decision process for that domain or that delimit the discussion around domain products, options, and standards. This content builds on the charter and assumptions of the overall process but also provides greater definition to the roles and responsibility of the domain team.
- Domain technology grid — this framework includes all of the domain technologies in place within the enterprise or planned for acquisition, including the following:
  - Technical description of the current state*

---

* The technical description should be kept to one page. Otherwise, colleagues will not read it. Basic information on this sheet should include product name and version number, the vendor, the domain team responsible for the maintenance of the entry, a brief description of the product, a rationale for the choice of this particular product, a statement on the practical useful life of the product, references to more detailed product information, and, most importantly, the individual or department currently responsible for the care and feeding of the product. See *The Hands-On Project Office*, http://www.crcpress.com/e_products/downloads/download.asp?cat_no=AU1991, chpt8~1~architecture Web site~example for a complete set of examples of domain team output.

- Transition strategies and sunsetting plans
- Near-term migration targets (two-to-three year timeframe)
- Underlying standards

■ Supporting documentation — this may come in many forms: vendor publications, consultant or in-house generated research, requests for proposals (RFPs), bibliographies, URLs, etc.

To bring this information together and to give it focus, the support team created a Web site organized so that each technology domain would have its own work space and tool set. This site promoted collaboration and afforded an ideal venue for communicating process results to the greater NEF/IS organization. See Exhibit 4.

The list of process deliverables comprehends all of the critical information required for action. For example, if NEF/IS managers want to know what particular technologies are in place within the firm today, all they need do is drill down on the Web site to uncover this information. If a particular systems development team needs to know what is in place to enable a particular business process and who within NEF/IS might be responsible for this infrastructure, that team also can drill down on the

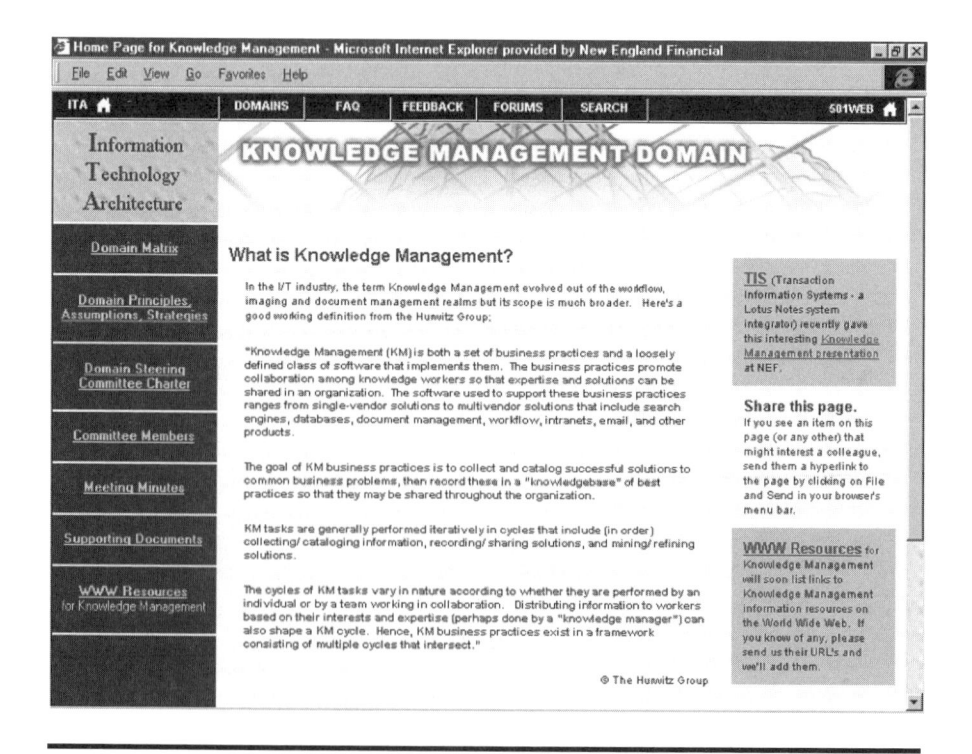

Exhibit 4   An IT Architecture Domain Team Home Page

Web site for this knowledge. When IT management prepares plans for the coming year's IT investments or if it is evaluating a particular technology for inclusion in the NEF complex, the architecture process knowledge store will provide a foundation for informed deliberations. In the case of the NEF/IS site, the domain teams tapped the very best minds within the organization for contributions. Similarly, we included contributions from our external partner providers to supplement internal expertise.

Throughout this process, we made it clear that the domain teams were not responsible for the actual planning and implementation of new technology investments. Rather, they were responsible for providing an informed direction and for the articulation of standards. Their recommendations served thereafter as the foundation for action by IT management, in line with the business needs and financial constraints of the enterprise. Many times, those involved in the architectural process also wear an operational hat. This was certainly true at NEF/IS, where domain leaders also oversaw IT teams and key lines of business. If this is the case, they must do what they can to keep these roles distinct; otherwise, the architecture effort will get mired in the minutiae of project implementation and management issues.

Each domain team, with encouragement from the PMO support team, began the ongoing process of gathering information and converting tacit knowledge into explicit architecture documents. The frameworks provided within the architecture Web site helped this effort along. See Exhibit 5.

The information grid illustrated in Exhibit 5 lays out the status of each technology within a domain category. For my example, I have chosen a view from the desktop and collaboration technology domain. To the far left, the domain team lists each technology product grouping, e.g., Microsoft Office. The team then indicates the minimum requirements for the product in question, as well as the enterprise standard. In the case of NEF, because the home office would often run different products than the field agencies, the form provides space for both options. The grid also allows space for listing product versions under study (i.e., In R&D). Products scheduled to be sunsetted or transitioned are also listed, as are targeted (future) standards. Product citations that are underlined indicate that they are linked to detailed product descriptions and other, more technical information. See Exhibit 6.

As part of the ongoing maintenance of an architecture process, the PMO team worked with the domain teams to ensure the currency and comprehensiveness of their respective technology grids and the linked product information sheets. This sort of detail management work aligns nicely with the competencies of a PMO. The author's support team also catalyzed domain team discussion by introducing information about emerging technologies and evolving business requirements. These efforts

Technology Component Category	INVENTORY of Standard and Tactical Technologies	TRANSITION PLANS for Legacy and New Technologies	DESIRED STATE (Emerging Technology up to 2-3 years out)	Underlying I/T Standards and Strategic Vendors
Office Suite *(click here to visit the Office support web site!)*	Minimum Supported Platform: MSOffice4.x and MSOfficePro4.x Current Home Office Standard: MSOffice97 and MSOfficePro97 Current Field Standard: MSOffice4.x and MSOfficePro4.x In R&D: MSOffice2000 and MSOfficePro2000	Sunset: MSOffice4.x and MSOfficePro4.x: 4Q99 Transition:	MSOffice2000 and OfficePro2000	Microsoft offerings
Web Publishing ("low-end")	Minimum Supported Platform:	Sunset:	Web publishing integrated into Office	Microsoft offerings

Exhibit 5 An Architecture Domain Information Grid

took time, but within three or four months, key domain team deliverables began to fall into place. Typically IT architect of the process assisted domain teams with grid extensions and on the initiation of applied research efforts. Lastly, as the body of information grew, the architect performed an audit function, keeping the domain teams and IT management apprised of the status and completeness of process deliverables.* In turn, these accomplishments helped shape the procurement, sunsetting, and standards decisions of the NEF/IS organization.

PUTTING THE ARCHITECTURE PROCESS TO WORK — IT PLANNING AND PROCUREMENT

Clearly, an architecture process requires a significant investment of time and effort. Given the up-front and ongoing cost of the work described here, how does one ensure the greatest benefit to the organization? One

---

* For an example of the standard NEF/IS architecture process audit, see *The Hands-On Project Office*, http://www.crcpress.com/e_products/downloads/download.asp?cat_no=AU1991, chpt8~2~architecture audit~example.

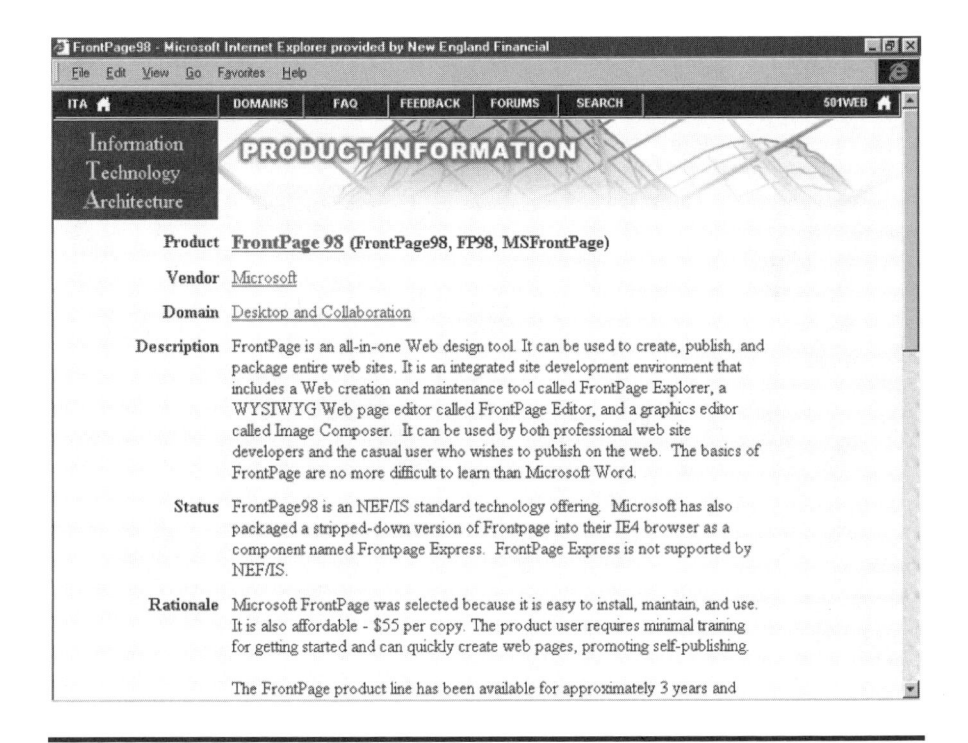

```
FrontPage98 - Microsoft Internet Explorer provided by New England Financial
File  Edit  View  Go  Favorites  Help
ITA         DOMAINS    FAQ    FEEDBACK    FORUMS    SEARCH              501WEB

Information
Technology          PRODUCT INFORMATION
Architecture

     Product  FrontPage 98 (FrontPage98, FP98, MSFrontPage)
      Vendor  Microsoft
      Domain  Desktop and Collaboration
 Description  FrontPage is an all-in-one Web design tool. It can be used to create, publish, and
              package entire web sites. It is an integrated site development environment that
              includes a Web creation and maintenance tool called FrontPage Explorer, a
              WYSIWYG Web page editor called FrontPage Editor, and a graphics editor
              called Image Composer. It can be used by both professional web site
              developers and the casual user who wishes to publish on the web. The basics of
              FrontPage are no more difficult to learn than Microsoft Word.
      Status  FrontPage98 is an NEF/IS standard technology offering. Microsoft has also
              packaged a stripped-down version of Frontpage into their IE4 browser as a
              component named Frontpage Express. FrontPage Express is not supported by
              NEF/IS.
   Rationale  Microsoft FrontPage was selected because it is easy to install, maintain, and use.
              It is also affordable - $55 per copy. The product user requires minimal training
              for getting started and can quickly create web pages, promoting self-publishing.

              The FrontPage product line has been available for approximately 3 years and
```

Exhibit 6   A Product Information Sheet

of the best vehicles for sharing and leveraging the accumulated knowledge of the architecture process is the enterprise's intranet. Through an architecture Web site, the architecture team can organize and present process findings attractively and economically. Each domain team can publish its updates to the site, keeping the knowledge base current. Most importantly, through the effective use of hypertext, document tagging, and end-user navigation design, the site's technology grid and other means of access can link numerous internal and external information sources, leaving the depth of data mining to the user. See Exhibit 7.

As depicted here, the NEF/IS architecture site home page readily conveys the domain structure of the process and affords easy access to domain-specific knowledge. This particular site also serves to inform and promote new IT standards, products, and services to the rest of the NEF/IS team. Throughout, the tone is positive and tailored to its technology audience. This Web approach also simplifies data collection and distribution. There is no need to create vast amounts of paper documentation or to limit oneself to any particular presentation medium. Furthermore, because many architectural elements will be supplemented with content from external, third-party providers, a Web approach affords easy integration among Web sites for added technical and product information.

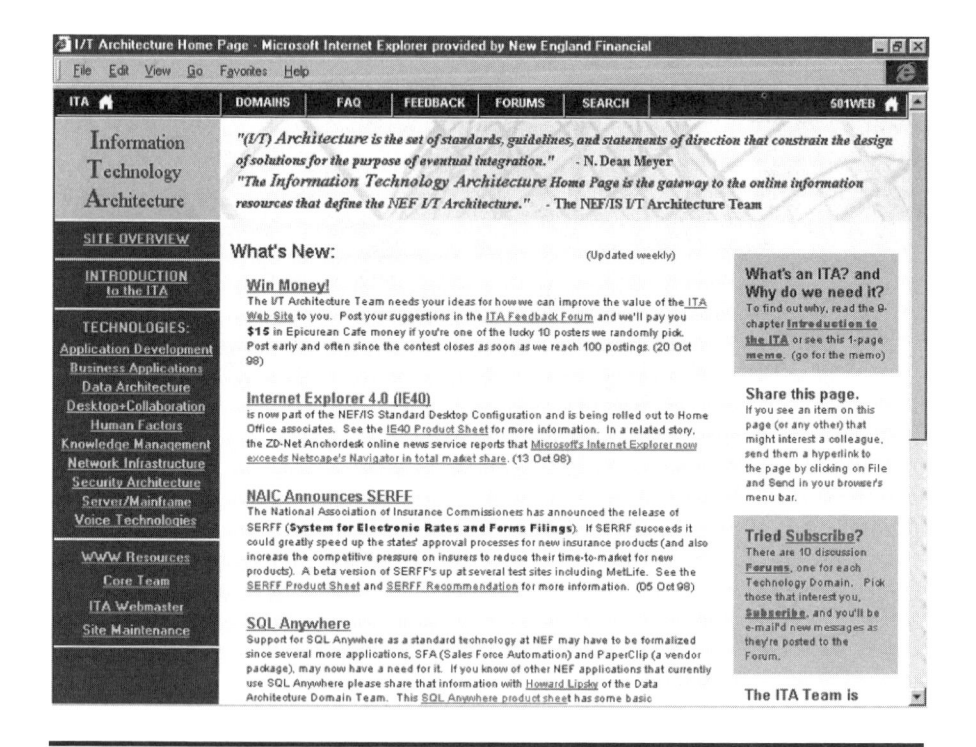

Exhibit 7 The NEF/IS Architecture Site Home Page

Thus, much of the detailed data collection may be left to others. Just create the links from the enterprise's intranet across the firewall to external Internet sites. As new technologies emerge and as the enterprise invests in different IT products, these changes may be readily reflected on the intranet site through domain team self-publishing (i.e., adding to the technology grid and the library of product information pages and links). The result is a single, highly accessible, and current source for all architecture-related information.

Beyond serving as a data repository and one-stop reference library, an architecture Web site may also afford a venue for cross-functional collaboration. Through domain-based discussion forums, those technologists who do not formally serve on domain teams have an opportunity to comment and expand on architecture contents. In so doing, they may also influence the outcome of enterprise technology selection. See Exhibit 8.

By employing a series of subscription-based discussion groups, interested parties will receive e-mail notices of any discussion forum updates. Through these forums, the greater NEF/IS community got involved in the architecture process, enhancing content, offering insights, and more generally buying into the knowledge sharing and standards compliance efforts. By ensuring the overall quality and comprehensiveness of the site, process participants

Architecting Success ■ 217

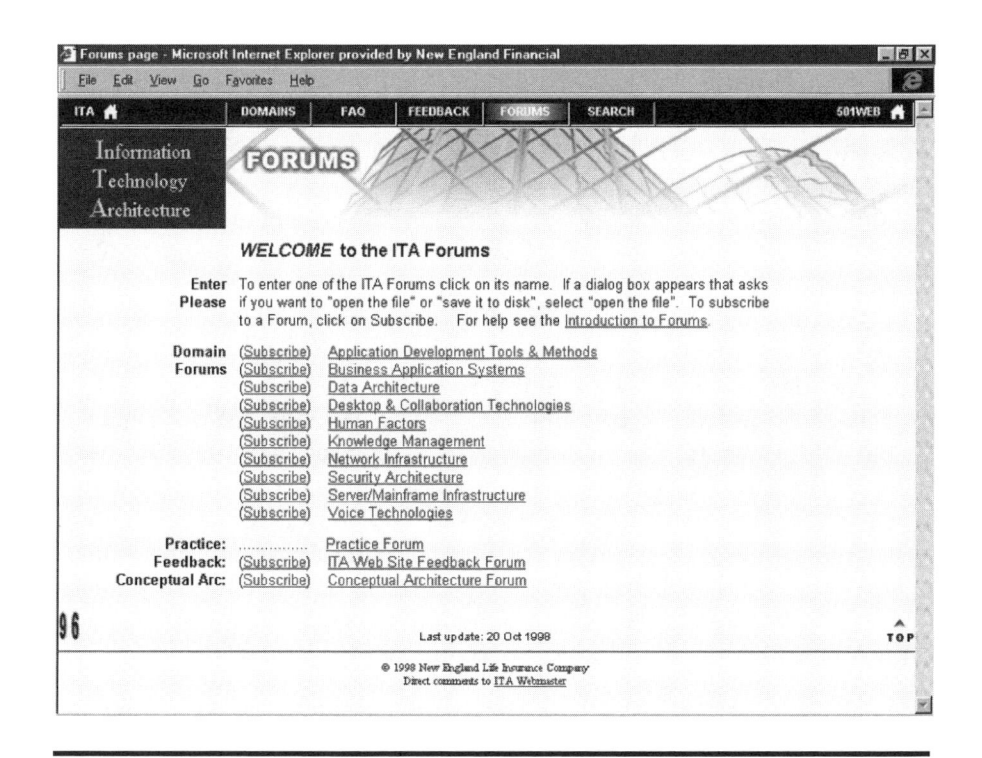

Exhibit 8   The NEF/IS Architecture Process Forums Structure

draw in operational users looking for background information on existing systems, development tool options, and strategic technology directions. For most of the NEF/IS organization, this was the first time that they had enjoyed access to such extensive, current information. Indeed, one of the most immediate and heavily used aspects of the Web site was the product sheets that identified particular IT product and service owners/experts.

If the first measure of an architecture process's impact is the extent to which it becomes commonly used within the IT organization, the advent of a well defined and easily navigated NEF/IS site succeeded in engaging our systems developers and technology integrators. They began to employ it as a reference tool. Over time, this use translated into less time wasted searching for technical information, the leveraging and occasional reuse of technology components, and a greater adherence to IT standards and practices.* These

---

* IT organizations that adhere to a software engineering framework and related disciplines will find that this architecture process nicely complements other process improvement efforts. When best engineering practices are married with sound architectural design standards, there are typically more predictable and higher-quality outcomes to systems development and integration efforts. If this is the case, it makes sense to give the oversight of the architecture Web site to the PMO.

consequences contributed in turn to lowering the cost of developing and maintaining IT products and services. Although NEF's lines of business still drove the purchase or in-house development of applications, with the architecture site in place, NEF/IS was positioned to advise them on their IT choices in a more effective and timely manner.

Although all this value may be appealing, the communication and use of an IT architecture framework is but one of its major outcomes. Two other aspects of this work should be emphasized, namely its capacity for auditing and for advocacy. In terms of the audit function, it is not enough for the process architect or the PMO to support passively the distribution of information to the IT community. These process staffers must continually monitor adherence to the standards articulated in the architecture. To this end, the author recommends a two-tier approach to auditing. On the one hand, the process architect should meet regularly with operations and systems development teams to ensure that there is an understanding of and general agreement with the organization's IT architecture framework. On the other hand, the architect should hold formal quarterly reviews of the enterprise's architectural compliance and communicate audit findings to IT organization management. See Exhibit 9.

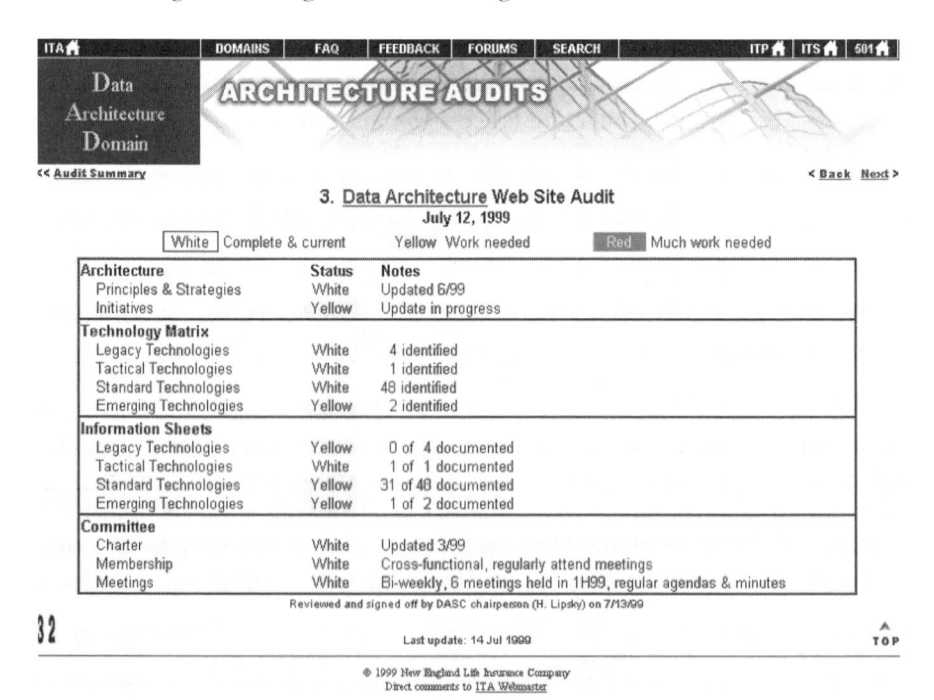

3. Data Architecture Web Site Audit
July 12, 1999

		White Complete & current	Yellow Work needed	Red Much work needed

Architecture	Status	Notes
Principles & Strategies	White	Updated 6/99
Initiatives	Yellow	Update in progress
Technology Matrix		
Legacy Technologies	White	4 identified
Tactical Technologies	White	1 identified
Standard Technologies	White	48 identified
Emerging Technologies	Yellow	2 identified
Information Sheets		
Legacy Technologies	Yellow	0 of 4 documented
Tactical Technologies	White	1 of 1 documented
Standard Technologies	Yellow	31 of 48 documented
Emerging Technologies	Yellow	1 of 2 documented
Committee		
Charter	White	Updated 3/99
Membership	White	Cross-functional, regularly attend meetings
Meetings	White	Bi-weekly, 6 meetings held in 1H99, regular agendas & minutes

Reviewed and signed off by DASC chairperson (H. Lipsky) on 7/13/99

Last update: 14 Jul 1999

© 1999 New England Life Insurance Company
Direct comments to ITA Webmaster

Exhibit 9 An IT Architecture Audit Summary View

Here again, the Web site may facilitate information gathering and sharing, as well as encourage compliance. The tool as employed at NEF/IS captures the following items:

- Status of the domain team's operating principles and frameworks
- Completeness of the domain's information grid, referred to in the example as the technology matrix
- Completeness of the individual product and service information sheets that flesh out the information grid
- Status of the team as an ongoing operating unit

Each domain team report indicates the work completed to date (white), work under way but not yet complete (yellow), and work that is stalled (red). In the spirit of continuous improvement, the support team helps its domain team colleagues identify those areas where either the architecture itself is lacking or where there is a need for greater effort by the domain team in defining and establishing IT standards. Here, you may wish to involve your IT organization's executive leadership more directly in setting a direction and ensuring sufficient time and focus for the process in support of your domain teams.

By monitoring compliance, the architect or the PMO educates the greater IT team about the gaps in IT architecture standards and knowledge while continuing to expand on and improve that foundation for informed action. Of course, not all the news that comes out of architecture audits will be well received, especially by those cast in a poor light. Auditors are never the favorites of those who fail to live up to their obligations. If the rules of the game are set out clearly in advance, however, and if all agree to play by those rules, the outcomes of audits should merely confirm what is already known. Furthermore, the findings of these audits may encourage additional attention to and perhaps resources for those areas of the architecture discovery process that have fallen behind schedule or that cannot keep up with the growing body of knowledge pertaining to their area of focus. In general, the audits at NEF/IS were received in this manner and contributed to the long-term strengthening of the overall process.

Lastly and perhaps most importantly, the architecture process affords a platform for the advocacy of best practices, for the disciplined life-cycle management of products, and for the introduction of new technologies. The architecture framework is a compendium of best-in-class solutions that marry with the enterprise's business requirements. If a particular solution has demonstrated its worth to the organization, these results should be championed by the architect and the appropriate domain committee. A complementary software engineering process, if applied across IT, will offer yet another venue for the architect and the PMO to advocate for the use of a particular technology across many different settings. Any occurrence along these lines reinforces the value of the organization's ROI on its architecture process investment.

Similarly, once the architecture has established an IT standard, management will possess a metric for assessing technology choices that are ultimately tied to ROI and the expertise of the IT organization. For example, most organizations have a difficult time retiring technologies. With an architecture knowledge platform, as illustrated in this chapter, IT management will be better positioned to establish expiration dates for enterprise hardware and software and to set business rules for the sunsetting of particular system components. This takes some of the political risk out of the effort for those directly involved, by promoting change in line with the overall direction of enterprise IT strategy and the application of accepted standards of measure. In fact, the author can point to several occasions when the parent of NEF, MetLife, employed information drawn from the architecture knowledge store to support system consolidations and other cost-saving measures.

Finally, the tool created by the IT architecture process will position participants to influence the strategic direction of technology investments. Through the applied research stored on the site and the associated forum discussions that these findings will stimulate, the process may be employed to identify specific choices and to build a strong consensus in support of these choices. Nor must these decisions emanate from the top of the IT organization. As the opportunities suggested by IT innovations become apparent to those serving on domain teams, these participants are ideally placed to develop and share ideas in line with the enterprise's business needs and the IT operational requirements. Through Web site forum discussions and associated applied research, refined versions of these bottom-up proposals may then be placed before management for formal consideration and action if appropriate. As the author witnessed at NEF, when proposals for change evolved out of architecture process activities, these plans enjoyed broad backing from NEF/IS technology thought leaders and service delivery managers. For that matter, because they were always framed within the context of document standards and best practices, they also tended to fit well with the rest of NEF's IT infrastructure.

CONCLUSIONS AND LESSONS LEARNED

In the final analysis, an IT architecture process is more about structure and discipline than it is about breakthrough thinking and radical change. Yet, one practice may lead to the other. To reap the benefits of the architectural process, your team must gather a vast amount of information, present it systematically, maintain its currency, and win over the greater

IT community to its application as part of business as usual. To achieve these tangible results, the process must involve a significant cross section of IT personnel from the outset — mostly the organization's best and brightest. The value inherent in the undertaking is multifaceted:

- Direct and tangible linkage between the enterprise's business strategy and its IT investment strategy
- Coherent, comprehensive, and collective IT vision for the organization (for both business and technology leaders)
- Enhanced ability of the IT organization to deliver quality products and services while controlling and even reducing the cost of their creation and ongoing maintenance
- Leveraging of the collective knowledge of the enterprise's IT assets and a more long-term view of asset value
- Enterprisewide, cross-functional consideration of IT resources
- Avoidance of unnecessary and uncoordinated investment in inappropriate technologies and IT research
- Regular assessments by the organization's IT leadership of its embedded base of technologies against external opportunities, threats, and business imperatives

But these benefits will only come to fruition if the architecture process delivers on its promises. To that end, your IT organization must participate directly in the formulation and maintenance of process deliverables and abide by its proffered direction. Although the business benefits detailed here may provide sufficient motivation, there are other justifications that may resonate more readily with your team's IT professionals. These benefits may be summarized as follows:

- Clear and uniform process for selecting and prioritizing IT initiatives
- Inclusive process that draws upon the expertise of the entire IT team
- Creation of an accessible repository for current-state technical documentation and future-state plans and directions
- Enforceable process for the sunsetting of superannuated technologies, allowing the technology team to focus on the most current products and services
- Accountable process for technology planning, change, and maintenance
- Process that shields the technology team from working with inappropriate technologies and from conducting unnecessary applied research
- More proactive approach to IT planning and risk management as these link to the enterprise's business needs and plans

Together, these two sets of drivers, one business-oriented and the other technology and engineering–oriented, should help IT executive management win support for and participation in the IT architecture process. As the NEF case study demonstrates, the IS organization took on an architectural process primarily for the second set of reasons. People in IT were tired of investing their limited time and energy in projects that did not go anywhere because these projects were so misaligned with business needs or based on inappropriate technology choices. Also, they were frustrated by working to support systems that were being maintained well beyond their useful life. Finally, they thirsted for access to leading-edge technologies and a voice in the selection and deployment of leading-edge products. The process, as described, helped them to realize these objectives and to mature as a better informed and more collaborative IT team.

For the business side of the house and for IT executive management, the ROI calculation involved different measures. As one might expect, their commitment was driven by the desire to rationalize the enterprise's technology platform, to consolidate system choices and associated support costs, and to reduce the overall total cost of IT ownership. It only became apparent subsequently that their investment in an IT architecture process led to more successful and more rapidly deployed information system solutions. Here, the ROI was less easily quantified but was appreciated all the same by NEF leaders. In the final analysis, the effort was vindicated by the parent company itself, because MetLife adopted and then adapted the process begun at New England Financial for its own uses.

For a modest investment, the IT architecture process will provide a valuable return. At the very least, it will establish for your enterprise a firm basis for making immediate investments and for addressing longer-term IT planning issues. However, the success of such an undertaking is not assured. First and foremost, it calls for a collegial form of leadership and facilitation, one that can comprehend the full scope of IT's technology investments without encroaching on the respective responsibilities of service and project delivery line managers. Second, effective architecture planning and documentation are all about process discipline, something that may run contrary to the dispositions of some IT professionals. Third, the effort must run as a KM undertaking: building communities of interest, winning the participation of thought leaders, organizing and managing large quantities of explicit information, and establishing ways to draw out and document tacit information.

Even a process architect brings only some skills, mostly around the integration of technologies, to the table. This person must partner with colleagues who possess a sense of the scope of the enterprise's technology investments and a solid understanding of how to get the job done. One of the obvious sources of such expertise is the PMO. Here, service and project delivery management responsibilities go hand-in-hand with an attention to process detail and a commitment to KM. By working together, technical architects and the PMO can bring to bear the skills required to support an IT architecture process. In doing so, PMO personnel will complement their other roles and responsibilities within the greater IT organization.

9

CONCLUSIONS — THE ROI OF THE PMO

INTRODUCTION

The Hands-On Project Office has taken the reader on a journey of exploration and self-examination. Throughout, the text has focused on service and project management delivery within an IT organization and how simple process changes can lead to higher customer satisfaction. To the author's way of thinking, a lack of clarity around business processes, stakeholder communications, and customer expectation management, rather than technological failures, leads IT operating units and those they serve down unhappy, sometimes torturous paths. The IT project management office (PMO) support team can address many of these issues by attending to the details of delivery management. To that end, this book and its accompanying tool set provide insights into the hows, whys, and wherefores of the PMO service and business model. In doing so, the book offers a starting point for the reconsideration of those methods and practices vital to the success of an IT organization.

To provide the proper context for these deliberations, Chapter 1 established an internal economy model for IT organizations, including a perspective on their operational, organizational, financial, and human resource dimensions. Chapter 2 discussed the roles and responsibilities within typical IT service and project delivery processes, as well as critical factors in their ability to deliver. This chapter then considered the likely benefits of creating a working PMO or its virtual services delivery group alternative. In short, the author can support a working model in which PMO-like activities are factored into the roles of IT management in lieu of an institutionalized solution, but such an arrangement may not function as well as an independent, objective, and highly focused PMO. If Chapter

2 focused on doing things right, Chapter 3 explored doing the right things, outlining a practical process and tool set for determining and pursuing IT investments of the highest priority to the enterprise. Here too, I carved out an important support role for the PMO in achieving desired outcomes.

The next three chapters dealt with the core services of any PMO: service delivery management, project management, and business process analysis and documentation. First, Chapter 4 examined service delivery management best practices, including the design and implementation of SLAs, the measurement of service level performance, the maintenance of reporting processes, and the more general support of the IT customer relationship management (CRM). Next, Chapter 5 took a focused approach to project management best practices, encompassing project scoping and commitment making, risk and resource management, the oversight of day-to-day project engineering and delivery processes, delivery measurement and reporting processes, and the more general coordination of IT project activity at the business unit portfolio and enterprise levels. Lastly, Chapter 6 offered a comprehensive tool set for discovering and documenting customer needs in ways comprehensible to nontechnical users but highly relevant for those assigned to build, operate, and service IT systems. For each of these three functional areas, I have married years of practical experience with simple, easily implemented, and proven tools that offer a starting point for the reconsideration of established processes. I also demonstrated the vital role that PMO personnel can and should play in the delivery of these services and process disciplines.

Chapter 7 and Chapter 8 offered two case studies that illustrate the significant benefits of knowledge management and Web-enabled collaboration in enhancing IT organizational performance. Chapter 7 demonstrated how an IT organization can promote best practices and the leveraging of IT community knowledge to improve team productivity and service and project delivery. As the creators or curators of the organization's process documentation, the PMO team is best situated to provide this valuable and complex KM function. As objective third parties to delivery, PMO personnel also enjoy a better sense of the opportunities for reuse and leverage within this vast array of documentation.

Similarly, Chapter 8 considered the PMO's support role in a collaborative IT architecture planning and asset management process. Because they are directly involved in chronicling all strategic project work, PMO personnel will be among the first to learn of changes in the organization's technical direction and how these decisions might influence current practices and the embedded base of existing IT. Furthermore, because they tend to be more aware of technical needs across the IT organization, PMO teams also have the right perspective to alert and prepare those parties most likely to exploit new IT investments as these come along. As in the prior KM case study, the PMO team is not a passive observer but works directly across IT to catalyze the reuse and repurposing of artifacts and corporate knowledge.

These illustrations, models, and case studies offer a lens through which the reader may compare and contrast a set of real-life scenarios and proven recommendations with his or her IT organization's own practices. Although *The Hands-On Project Office* offers no absolute truths, I hope that my ideas may be adapted and applied to address your team's particular management needs. In some instances, you may be obliged to readjust my frameworks to make them more in keeping with the context of your business. Others will work for you right out of the box. At the very least, you will find any number of useful ways to assess the current effectiveness of your IT delivery teams and the ROI for looming IT investment opportunities. Most importantly, the case made here for a PMO asks you to consider which aspects of these models and tools may be implemented for greatest impact in your own IT shop.

Once you have recognized and internalized the value proposition for a PMO, you must find a clear and cogent way to sell this idea to two tough constituencies. On the one hand, you will be faced with convincing enterprise leaders (i.e., your potential sponsors) who will view the expenditure on a PMO as merely an increase in overhead. Here, you must make a clear link between the deliverables of the PMO and an overall improvement in IT team cost and performance. On the other hand, you may find that your own IT staff objects to what they perceive as encroachments on their own areas of responsibility. Here, you must demonstrate that the work of the PMO complements and does not subvert their efforts. Furthermore, you must show that its portfolio of value-added services closes the gaps among existing IT practices while enhancing the resources allocated to particular aspects of IT work.

To be sure, neither sale will come easily. All of these concerns have validity and should be addressed before the rollout of a formal PMO implementation. Therefore, it is essential that the IT leadership makes a strong case for the investment. This chapter considers an array of arguments to win stakeholder support for launching and maintaining an IT PMO, beginning by reframing some of the questions that opened *The Hands-On Project Office* and then revisiting the ROI model featured in Chapter 2.[*] As before, tailor the frameworks that follow to fit the particular circumstances of your business setting.

---

[*] The complete PMO ROI analysis tool may be found at *The Hands-On Project Office*, http://www.crcpress.com/e_products/downloads/download.asp?cat_no=AU1991, chpt2~9~PMO value calculation~model and template.

THE ROI DISCUSSION

In preparing your team for the launch of an IT PMO initiative, you might begin with a little gap analysis of your own. The dimensions of this inquiry should embrace particular points of performance weakness within the IT organization, as well as areas of practice, such as business process reengineering knowledge, management, or architectural planning, where the existing organizational roles and responsibilities of your group preclude a consideration of such staff functions. Here are some of the questions you might ask yourself and your management team:

- What is the state of our current IT planning process?
- Is the allocation of IT organization resources properly aligned with enterprise priorities?
- If there is an IT planning process in place, what percentage of IT management time is devoted to that process, and is this time and effort well spent?
- How are IT's relations with our key customers (i.e., sponsors)? Do we and they understand one another, speak the same language, and agree on similar priorities?
- How does our organization measure and report on its performance to those who fund our activities?
- Are our customers satisfied with our service delivery? How do we know?
- Are we managing customer expectations and, if not, what does it cost us every time we disappoint a sponsor?
- How many resources do we currently devote to reactive problem correction (e.g., call center, CREs, maintenance and support personnel)?
- Are our customers satisfied with our project delivery? Do we deliver our projects on time, in budget, and in keeping with customer requirements?
- How often are IT resources misdirected because of a misunderstanding about customer specifications, project change orders, or performance metrics or reports?
- How much IT staff time is devoted to finding out who has particular expertise, manages particular IT assets, or holds the responsibility for a particular area of service or project delivery?
- What skills and experiences are needed for our team's success? How are these requirements captured and addressed today?
- Does our enterprise suffer from the proliferation of information technologies or do we promote architected IT solutions built around recognized standards? What is the total cost of IT solution ownership?

Through such honest introspection, you will quickly come to understand those shortfalls in focus, resource allocation, or competence that expose your organization to service and project delivery failure. For example, if your self-assessment leads to the conclusion that the IT organization is poorly aligned with the enterprise's business priorities, this calls into question the appropriateness of your unit's investment decisions. If communications and relations between your delivery teams and their customers are frayed, even favorable service and project delivery outcomes might not be received as such. Similarly, if your outcomes are plagued with significant scrap and rework, delivery delays, and excessive staff overtime, this also would suggest a lack of alignment, poor communications, and an absence of effective delivery management practices.

Unfortunately, any number of IT organizations suffer from these shortcomings, not because their people are incompetent but because their leaders expect technology line managers to embrace these additional duties. As I have already suggested, these assignments are often a mismatch of skills, interests, and personal makeup of line personnel. Instead, I recommend that you employ a PMO team to complement your technical managers. To that end, focus the efforts of the PMO in support of team delivery, interpersonal communication, and business process needs of technologist colleagues. The particular PMO roles encompassed in such an approach might include the following:

- Resource conservation — assuming the support tasks associated with strategic planning, CRM, competitive benchmarking, process reengineering, documentation, reporting, and the like, to free IT executives and their overtaxed IT line managers to focus on those areas of strategy and service and project delivery where they will have the greatest impact
- Risk management — serving as the objective third party to and the documenters of any and all IT ventures to ensure that those directing projects are fully cognizant of the risks associated with those undertakings; developing and implementing approaches to risk mitigation
- Accountability and CRM — maintaining customer relationship portfolios and otherwise keeping the IT organization's CREs staffed with materials and information so that they, in turn, may interact with IT's executive customers (sponsors and working clients) in an effective and efficient manner
- Doing the right things — documenting and communicating IT organization priorities in line with the unit's alignment with and commitments to its line-of-business sponsors

- Doing things right — championing best practices, disciplined processes, performance metrics, accountability reporting, and lessons-learned activities in line with service and project delivery
- Total cost of IT ownership — enabling the streamlining of technology platform choices and the standardization of IT deployments through applied research, objective analysis, comprehensive financial modeling, and the honest assessment of the organization's investment strategies and spending practices
- Internal harmony, communication, and collaboration — facilitating communication and collaboration among IT delivery teams and between those teams and their customers

The need for these PMO services is borne out each time an IT team fails in its service and project delivery efforts. With a PMO in place, the IT organization is better positioned to balance the dynamic pressures of the business for new or expanded products and services against the internal tensions of running complex and fragile IT operations with constrained resources. My PMO model adds strength and flexibility to the mix of your IT organization's capabilities by placing appropriately skilled people at the most likely points of failure within key IT processes. In doing so, your organization may derive any number of benefits:

- Increased revenues from the successful delivery of improved IT products and services
- Decreased costs from the rationalization of business processes and technology platforms, and from the reuse and repurposing of project artifacts, system components, and team knowledge
- Cost avoidance due to a reduction in scrap and rework and a better alignment between enterprise needs and IT investments
- Increased productivity within the IT organization itself thanks to process reengineering, and an adherence to technology standards and architected solution designs
- Improved time-to-market thanks to the rigorous adherence to industry best practices, reuse, repurposing, and the adoption of standards-based solutions
- Improved customer-servicing capabilities and an enhanced perception of value thanks to better communications and relations with customers
- Managed risks through adherence to the practices of project life-cycle management, rigorous project planning, performance measurement, and a regular, open review process

These accomplishments establish a framework for the overall improvement of an IT organization's delivery capabilities. Your business colleagues will surely value contributions that lower costs and increase revenues. They will also appreciate on-time, in-budget project results that lead to improvements in line-of-business performance. They may even recognize the benefits of risk mitigation and of compliance with industry standards. Other returns from the investment in a PMO, however, may be less perceptible. For example, the office's work in enabling an effective KM practice may contribute significantly to IT's time-to-market delivery, but the arcane work of building a knowledge store may appear, at first blush, of little value. Similarly, the applied research and discussions that will lead to architected system solutions may not carry a clear ROI, even though such work may mitigate escalating costs for downstream application maintenance and integration.

Solutions like these will undoubtedly contribute to the long-term success of IT. But how can these be presented in such a way that they appear as something other than overhead costs in the thinking of nontechnical observers? Because technical arguments alone will not suffice, IT leadership must convey these benefits in ways that are meaningful to business colleagues. To assist the reader in this important exercise, the author would like to revisit my ROI model and offer a few additional insights into how this may assist in building a case for a PMO within the IT organization.

EXECUTIVE SUPPORT SERVICES

In the first grouping of ROI metrics, my model considers those PMO staff-support activities grouped under the broad heading of executive support services. Through the efforts categorized in the tool, including process design and documentation, business requirements gathering, competitive benchmarking, and facilitation, PMO personnel assist IT management through a sustained planning effort. As described in Chapter 3, this process is highly detail oriented and iterative. It requires a deep understanding of IT operations, a lot of data collection and analysis, and a clear sense of how best to communicate all of this content in an economical fashion to two very different audiences: line-of-business partners and the IT organization rank and file. See Exhibit 1.

Although the IT executive leadership will provide the direction for the planning process, and although line management will implement the plans once approved, some group must shepherd this process through its life cycle. For that matter, at its annual culmination, someone must ensure that the plan's commitments are properly conveyed to those responsible for delivery. Clearly, PMO personnel are well positioned for the facilita-

Exhibit 1 The PMO ROI — Executive Support Services

Value/Cost Categories	Amount of Financial Benefit/Non-PMO IT Costs	IT Costs Avoidance Associated with Service Delivery Risk Mitigation	IT Costs Avoidance Associated with Project Delivery Risk Mitigation	PMO Investments in Services and Risk Mitigation	Outcomes (Net Value of Positive Outcomes and Risk Avoidance Less PMO Costs)	Comments
Planning Process						
Requirements						
Benchmarking						
Metrics/targets						
Template preparation						
Process management						
Presentations						
Documents/updates						
Reporting						

tion/communication part of this process and, given the scope of their competencies and day-to-day responsibilities, are well suited to the front-end facilitation and analysis work, as well.

If a PMO is not in place to support these functions, the IT organization typically turns to one of three scenarios. First, it may choose to do no planning whatsoever. I will let the reader draw his or her own conclusions about the ROI of such inaction. Second, IT management may take on the planning process, which invariably means the process gets short shrift, remains at a high level of abstraction, and does not readily integrate with the day-to-day operations of the greater IT team. Third, IT management may turn to a corporate planning office or an external consulting firm for these services. Although the former option may be "free" to the IT organization and the latter rather costly, neither service provider would possess sufficient knowledge of the IT organization, its players, and its customers to ensure an end product likely to have value to the team or be implementable.

The PMO will provide executive support services at the lowest cost commensurate with a usable result. PMO staff will do so without putting an undue burden on IT management and will execute assignments as a complement to other PMO tasks. To get similar results through an external resource would be prohibitively expensive, and to employ generalists from within the enterprise, who may lack a deep understanding of IT, will usually require a considerable amount of IT management handholding. Thus, the value of the PMO to IT and the enterprise may be calculated in terms of the delivery of an essential service at incremental additional cost and without burdening the rest of IT management. Furthermore, the PMO is best positioned to link the final plan to its other work in support of IT service and project delivery teams.

SUPPORTING SERVICE DELIVERY

Few responsible enterprise managers would dispute the value of IT services. On the other hand, the reader may find little agreement within his or her business about what those services should be and how much line-of-business colleagues must pay for them. More often than not, the heroics of IT services providers on the shop floor and the small, day-to-day victories of IT-enabled business process change never reach the eyes and ears of management colleagues. As Chapter 4 explains, the disconnects in understanding occur throughout the delivery process, from initial expectation setting to an agreement on appropriate measures, and from the regular reporting of results to annual reviews of customer satisfaction.

High-level service relations (i.e., between IT and the leadership of an enterprise operating unit) should be handled by an IT executive. This person must be properly staffed to deal with his or her customer, including a detailed service history and issues for each account. Similarly, the modeling of performance measures, tracking of results, and compilation of monthly, quarterly, and annual reports are all necessary components of the service management process. Those on the line are too busy to carry out such work. Furthermore, customers are less likely to be frank with the party actually delivering the service under review. Here again, the process expertise of the PMO and its IT-wide perspective, as well as its separate organizational identity from that of service delivery teams, positions the PMO to provide such support.

From the standpoint of self-management, the mechanisms of PMO service delivery support will strengthen the sense of ownership and personal responsibility within your service teams. The mechanisms employed clearly link IT services to those served and detail the measures of customer satisfaction. By employing these metrics and sharing the results broadly within IT, the PMO reinforces IT's commitment to quality service delivery and calls out those who are letting down their side. When the process also embraces constructive responses to breakdowns in performance, such as internal process reengineering and the replacement of faulty hardware and software, these activities can transform the very culture of the IT organization.

Furthermore, customers will take note of these activities. Their first reaction will be that IT really cares about service delivery. This realization alone can do wonders for the relationship between IT and its business customers. As the process matures, opportunities will arise for engaging your end users in these continuous improvement efforts. For example, the data generated from metrics may suggest a need to invest in more customer training or in desktop system upgrades. Partnering in the area of service improvement will only strengthen the bond between IT and the lines of business, which in turn will make them more receptive to working with you on larger projects down the road. Here again, the PMO can play a critical role in the success of the process. Although the main actors may be your CREs on the one hand and your service delivery teams on the other, the PMO supports both constituencies. See Exhibit 2.

First, the PMO is positioned as a clearinghouse for service level information. This is a complex, highly detailed, cross-enterprise task. Your delivery teams will not have the time, the interest, the perspective, or perhaps the skills to do this work on their own. Second, the PMO enjoys enough distance from the services in question to identify and collect data on the true measures of customer satisfaction. Third, because in all likelihood it will manage both the historical knowledge base of activity and the forms and processes for SLA creation and delivery metrication,

Exhibit 2 The PMO ROI — Service Delivery

Value/Cost Categories	Amount of Financial Benefit/Non-PMO IT Costs	IT Costs Avoidance Associated with Service Delivery Risk Mitigation	IT Costs Avoidance Associated with Project Delivery Risk Mitigation	PMO Investments in Services and Risk Mitigation	Outcomes (Net Value of Positive Outcomes and Risk Avoidance Less PMO Costs)	Comments
Service Level Management						
Customer Group A						
Requirements						
Metrics/targets						
Template preparation						
Process management						
Documents/updates						
Reporting						

the PMO is positioned to act quickly and effectively in support of IT management.

Although service delivery management requires constant care and feeding by the PMO, the return on this investment is huge as measured in terms of the quality and strength of IT's relations with its customers and the coherence of the organization's service offerings as perceived by end users. Lastly, any consideration of the value of the PMO in this area must also recognize the substantial benefits its work will engender across IT. By highlighting the value chain in service delivery, the PMO will encourage better teamwork, collaboration, and heightened sensitivity to the needs of end users. Thus, by adding considerable rigor to the service delivery process, the PMO will strengthen internal IT operations and the relationship between the enterprise and IT service delivery teams.

SUPPORTING PROJECT DELIVERY

In terms of project delivery, the PMO's role is central. Without project management and business analysis services, few projects would ever meet their objectives. Although it is conceivable that the project director and other team members could take on these responsibilities, Chapter 5 and Chapter 6 outline why this is not an attractive option. In contrast, in establishing a PMO, the IT organization creates a focal point and eventually a center of excellence for those competencies of greatest importance to project administration. These contributions are best summarized in the ROI tool itself. See Exhibit 3.

Project management is all about risk management. At the strategic level and on behalf of IT management, the PMO can monitor the positioning and interdependencies among projects as these relate to overall IT organization commitments and enterprise IT priorities. The rest of the project team will not have the focus or the necessary perspective to do so. When developments in other projects may impact the project at hand, PMO personnel serve as an early warning system and can prevent others from making dangerous assumptions about the health of these dependent relationships. Similarly, the project manager works under the direction of the project director to ensure that the resources allocated to the project align with its commitments. The project manager is positioned to alert management should problems arise.

The project manager is also the guardian of project execution best practices. This person drafts and updates the project plan, holds team members accountable for their responsibilities, and reports on progress to IT management and customers. Typically, project directors are not particularly interested in these processes. Their focus is on getting the work done. But in their pursuit of results, team leaders may overlook

Exhibit 3 The PMO ROI — Project Delivery

Value/Cost Categories	Amount of Financial Benefit/Non-PMO IT Costs	IT Costs Avoidance Associated with Service Delivery Risk Mitigation	IT Costs Avoidance Associated with Project Delivery Risk Mitigation	PMO Investments in Services and Risk Mitigation	Outcomes (Net Value of Positive Outcomes and Risk Avoidance Less PMO Costs)	Comments
Project Delivery Management						
Project A						
Strategic risk						
Financial risk						
PM risk						
Technology risk						
Change management risk						
Quality risk						

many of the risks inherent in complex project work. The PMO's project manager balances this perspective with a commitment to process and with an understanding of how the experiences of other IT teams may benefit the current working group. In terms of technology risk management, the project manager's knowledge of the broad fabric of enterprise IT, IT organization standards, and the status of other IT projects that may serve as predecessors to the current project can leverage the work of others and mitigate technical risks.

In all of these ways, the PMO's project managers and analysts provide considerable value to project teams. In terms of the overall success of project delivery, however, perhaps the greatest contribution of PMO personnel is in the area of change management. Here, their focus is on ensuring that the customer is positioned to receive and use IT project deliverables once these are available. This work can take on any number of aspects. First, the PMO team will document and assess the business processes to be enabled through IT. Do these processes map to the envisioned IT solution? If not, what is to be done? Second, PMO staff will ask: Are business unit personnel prepared in terms of training, and are their IT platforms sized to exploit the new technology properly once it is in place? Third, the team will consider: Are related IT service delivery processes complementary to the new relationships and needs that will emerge from the introduction of this particular technology into the business? Those assigned by the PMO will address these questions and, in so doing, ensure positive outcomes from the introduction of IT across the enterprise. Not only are these deliverables critical to the success of IT project delivery, it can be argued that only individuals who possess the skills profile of a PMO member can carry out these critical assignments.

The overall PMO contribution in project management delivery revolves around the quality of process deliverables. PMO project managers and analysts ensure completeness through the IT development/delivery life cycle. By drawing on their past experiences and the PMO's project knowledge base, they can help their respective project teams avoid past mistakes and make the most out of each project outcome. They are there to mitigate the numerous risks associated with delivery and, when things do go south, they are properly positioned to capture and communicate what went wrong, why it happened, and how to avoid such failures in the future. Surely these benefits in and of themselves constitute a significant ROI on your investment in a PMO.

LEVERAGING TECHNICAL KNOWLEDGE

Perhaps the most quantifiable findings of the author's ROI model concern the PMO's impact on technology spending, reuse, and repurposing. See the achievements chronicled in Chapter 7 and Chapter 8, which reflect on the direct benefits stemming from the PMO's management of the IT organization's technical knowledge. Although IT leadership will make all of the decisions concerning standards and architectural direction, these cannot be achieved in a vacuum. PMO personnel generate the explicit knowledge required to document the current state of enterprise IT, and it is through their KM efforts that this information will be made accessible to the organization's management.

However, the role of the PMO in this regard need not be passive. Indeed, the PMO team should serve as the champion of technical standardization and the construction of architected IT solutions. Similarly, in their roles as project managers and analysts, PMO staff members should proactively leverage the IT unit's knowledge store in search of opportunities to reuse or repurpose existing work and deliverables. In short, they can contribute to cost savings and speedier time-to-market by applying these principles to the design and delivery of new IT solutions. See Exhibit 4.

On the process side of their assignments, PMO personnel will build tools, templates, checklists, and frameworks that save others steps and time in the execution of service and project work. Similarly, when they recognize connections between past projects and new ones, they can redeploy old projects' artifacts (e.g., project plans, budgets, business specifications, and commitment documentation) to jump-start more current efforts. They also can influence their project teams to employ standard architectural components (i.e., reuse) and to leverage off the work of colleagues (i.e., repurpose) rather than start over each time. Besides the advantages to the project in question, these practices will lead, over time, to a narrowing of the enterprise's technical base. Thus, by employing more extensively what is already in place, IT teams can deliver more promptly at lower costs and can reduce downstream maintenance and support exposures.

All of these benefits directly impact the business' bottom line. For its part, the PMO creates and maintains a KM environment that enables this approach. In doing so, the potential is there to avoid substantial costs while achieving more satisfying and sustainable results for the enterprise. To realize the value of this investment, PMO personnel must be in a position to influence those making IT architecture and technology acquisition choices for the enterprise.

STAFF SUPPORT AND IT ORGANIZATION CULTURE

The PMO can contribute to the overall health of the IT organization in any number of ways. Through its process stewardship, the PMO promotes

Exhibit 4 The PMO ROI — Architecture Management and Support

Value/Cost Categories	Amount of Financial Benefit/Non-PMO IT Costs	IT Costs Avoidance Associated with Service Delivery Risk Mitigation	IT Costs Avoidance Associated with Project Delivery Risk Mitigation	PMO Investments in Services and Risk Mitigation	Outcomes (Net Value of Positive Outcomes and Risk Avoidance Less PMO Costs)	Comments
Technology Platform Standardization and Rationalization						
Hardware costs						
Software costs						
Support costs						
Contractor costs						
Customer servicing costs						
Staff training costs						
Other costs (facilities, back-up, etc.)						

service and project delivery discipline but also a strong sense of joint ownership and responsibility for customer satisfaction within the greater IT team. Although PMO personnel monitor commitments and keep their colleagues accountable through measurement and reporting, these efforts also encourage a culture of continuous improvement through on-the-job learning and self-correction. If the PMO is established and promoted in this fashion, over time its practices will lead to a healthier, more collaborative work environment for all IT personnel. In turn, these developments will strengthen employee retention and reduce turnover, which are very tangible benefits to the organization. See Exhibit 5.

As the PMO builds the IT organization's knowledge store, it captures the histories of IT's successes and failures. It will also document all of IT's practices and standards. In doing so, the PMO creates the very materials needed to provide applied management and technical training to the rest of the IT organization. Rather than spending large sums on more generic staff training and development programs, the PMO team itself (or in conjunction with other corporate training experts) could provide much more focused and relevant offerings to the team at costs well below commercial rates.

Finally, the work of the PMO may be employed as the basis for IT employee evaluation. As part of its work regimen, the office captures individual and team accomplishments, as well as the associated measures of customer satisfaction. This data may be integrated with the unit's performance review process to link more clearly awards and recognition with the standards of best practice and delivery as promoted by IT management. Such a process demonstrates the IT organization's commitment to aligning what it does with the needs of the enterprise. These practices, in turn, send a strong positive message to your team and your customers alike. They are yet another example of how the work of the PMO enables the mission of its parent IT organization.

ONE LAST LOOK AT THE PMO

To conclude this book, the author returns once again to his snapshot view of the PMO value proposition. See Exhibit 6.* This graphic captures the sweep of PMO responsibilities in one integrated view. In applying my suggestions, the reader may find it necessary to recast this picture in his or her own terms. Nevertheless, I hope that the value of the overall message remains. To succeed as an IT organization, you must focus on

---

* For an electronic version of this model, see *The Hands-On Project Office*, http://www.crcpress.com/e_products/downloads/download.asp?cat_no=AU1991, chpt9~1~the PMO Value Proposition~model.

Exhibit 5   The PMO ROI — Staff Support

Value/Cost Categories	Amount of Financial Benefit/Non-PMO IT Costs	IT Costs Avoidance Associated with Service Delivery Risk Mitigation	IT Costs Avoidance Associated with Project Delivery Risk Mitigation	PMO Investments in Services and Risk Mitigation	Outcomes (Net Value of Positive Outcomes and Risk Avoidance Less PMO Costs)	Comments
Staffing Costs						
Management training						
Technical training						
Retention						
Recruiting						

Conclusions ■ 243

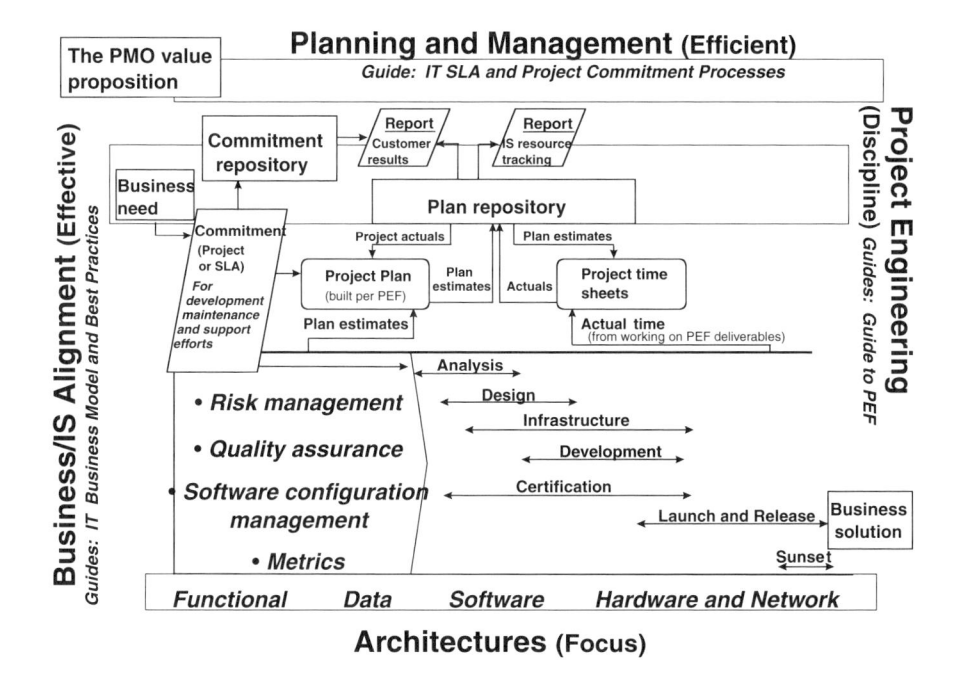

Exhibit 6 The Project Management Office — Overview and Value Proposition

excellence in service and project delivery. Although many factors will contribute to your success in these regards, for most enterprises the establishment of a PMO will help to ensure that success.

I have offered for your consideration stories of past experiences and examples of the methods and tools employed in the course of my work. In particular, I have examined those processes that focus on alignment of purpose, management of service and project delivery life cycles, measurement and reporting of results, and leveraging of corporate knowledge. The world of IT deployment and management is a tough and unsettling one in which to toil. If I have not entirely lifted the reader's burden, then perhaps I have suggested some useful ways to do a better job in trying circumstances.

APPENDICES

KEY TEMPLATES FROM THE PMO TOOL BOX

As part of its commitment to applied learning, *The Hands-On Project Office* provides a project management office (PMO) tool box. Found online at http://www.crcpress.com/e_products/downloads/download.asp?cat_no= AU1991, the complete eTool Box includes electronic versions of the templates, business models, and process workflows discussed in this book. In addition, you will find examples of the application of key tools in real-life IT management settings. For your convenience, some of the key tools also appear in the following appendices for easy reference when reviewing the text.

Appendix A — IT Project Justification Template 247
Appendix B — IT Annual Plan Template 253
Appendix C — PMO Value Calculation — Model and Template 263
Appendix D — Service Level Agreement Template.............................. 271
Appendix E — Project Management Life-Cycle Framework................. 283
Appendix F — Project Leadership Questionnaire for Change
 Management Projects 295
Appendix G — IT Project Risk Management Matrix............................. 301
Appendix H — Commitment Document Template 305
Appendix I — Master Project Schedule Template............................... 317
Appendix J — Glossary 321
Appendix K — Selected Readings 329

APPENDIX A:
IT PROJECT JUSTIFICATION TEMPLATE

Information Technology Project Request Form

Project Name:

Business Sponsor(s):

FYXXXX Funding Request:

Project Summary:

Project Deliverables/Milestone Dates:

Current Quarter:

Working Client(s):

Next Three Quarters:

Appendix A: IT Project Justification Template ■ 249

Summary of Costs:

Quarter	IT Headcount	IT Costs	Vendor Costs	Hardware/Software Costs	Other IT Costs
Current					
Next Qtr 1					
Next Qtr 2					
Next Qtr 3					
Next Qtr 4					

Project Roles and Head Count Distribution

Role	Description	Head Count
Working client(s)	Directly responsible to the business unit sponsor for project delivery, coordination, and facilitation	
Other business unit representatives	Those parties from across the business unit who participate in project delivery through their contributions of marketplace, customer, business process, and operational knowledge	
IT customer relationship executive (CRE)	The IT staff member who manages the relations between this particular customer and the entire IT organization	
Project director	The IT manager directly responsible for the delivery of this project for the IT side of the house	
Project manager	Staff to the project director who coordinates the day-to-day process of project delivery	
Business analyst(s)		
Technical architect		
Systems personnel		
Data management personnel		

Quality assurance/release
management personnel _____
Security _____
Server and network
services _____
Training _____
Other _____
Total _____

- What, if any, is IT head count currently associated with this initiative? _____
- Confidence level of estimates: Low___ Medium___ High___
- Explain the confidence rating: _____

Summary of Benefits:

- What is the associated IRR/ROI (attach the appropriate calculation model)?
- What were the benefits based on (i.e., expense reduction, increased sales, etc.)?
- Dollar amount of benefits realized in:

	FYXXXX	*FYXXXX+1*	*FYXXXX+2*	*FYXXXX+3*
1st Qtr				
2nd Qtr				
3rd Qtr				
4th Qtr				

Strategic Alignment (Y/N):

- Supports increased sales, market share, revenue, etc.:
- Reduces expenses:
- Supports demutualization:
- Legal/mandatory/compliance:
- Supports sales force retention:
- Supports IT architecture/blueprint:

Risk Factors (Y/N):

(Note: Mitigation plans must be identified for each risk.)

- Strategic risk (impact to business if application is not delivered, if it fails in production)?
- Financial risk (soundness of budget, security of funding, actuals in line with budget)?
- Project management risk (scope creep, cost/time overruns, people)?
- Technology risk (performance, deployment, support, integration/interoperability, standards, expertise/competencies)?
- Change management risk (change process documented and followed, business requirements stable)?
- Quality risk (requirements clear, project results traceable to requirements, reviews with customer in place)?

APPENDIX B:
IT ANNUAL PLAN TEMPLATE

IT Organization Action Plan, FYs 2xxx–2xxx

Information services vision statement	
Information services mission statement	
Information services goals and objectives	• Goal 1: • Goal 2: • Goal 3: • Goal 4: • Goal 5:
Information services competitive profiling	

Note: All commitments are based on a July–June fiscal year, thus 3rd Q 2003 = January–March 2003.

Appendix B: IT Annual Plan Template ■ 255

Information services guiding principles	A critical element of building a high-performance, service-oriented team within information services is to establish a number of cultural characteristics that permeate the organization. These guiding principles provide both the framework and the expectation for the maturing of the organization. • Principle 1: • Principle 2: • Principle 3: • Principle 4:
Information services critical success factors	A critical element in running IT as a business is setting clear operational characteristics regarding how our projects and services will be managed and delivered. As with the guiding principles for cultural characteristics, these critical success factors provide the framework and the expectation for our operational maturation. Factor 1: • • Factor 2: • • • Factor 3: • • • Factor 4: • • • •
Note on funding	
Performance and satisfaction measurement	

256 ■ The Hands-On Project Office

Strategic Goal # 1

Key Initiatives	Responsible Parties	Timetables/Resources	Performance Metrics	Self-Review	Manager's Review
A.	•	•	•	•	•
B.	•	•	•	•	•
C.	•	•	•	•	•
D.	•	•	•	•	•

Strategic Goal # 2

Key Initiatives	Responsible Parties	Timetables/Resources	Performance Metrics	Self-Review	Manager's Review
A.	•	•	•	•	•
B.	•	•	•	•	•
C.	•	•	•	•	•
D.	•	•	•	•	•

Strategic Goal # 3

Key Initiatives	Responsible Parties	Timetables/Resources	Performance Metrics	Self-Review	Manager's Review
A.	•	•	•	•	•
B.	•	•	•	•	•
C.	•	•	•	•	•
D.	•	•	•	•	•

Appendix B: IT Annual Plan Template ■ 259

Strategic Goal # 4

Key Initiatives	Responsible Parties	Timetables/Resources	Performance Metrics	Self-Review	Manager's Review
A.	•	•	•	•	•
B.	•	•	•	•	•
C.	•	•	•	•	•
D.	•	•	•	•	•

260 ■ The Hands-On Project Office

Strategic Goal # 5

Key Initiatives	Responsible Parties	Timetables/Resources	Performance Metrics	Self-Review	Manager's Review
A.	•	•	•	•	•
B.	•	•	•	•	•
C.	•	•	•	•	•
D.	•	•	•	•	•

Table Key:

The performance metrics section of this form refers to various measurement processes that are linked directly to the critical success factors for that action item. While the action items themselves are discrete deliverables in the overall IT plan, measurement mechanism will be more common. For example, the references to "customer satisfaction" concern a single tool administered to IT division customers about all aspects of IT operations. Thus, this plan envisions a capability whereby common tools will measure the performance of discrete products and services. As part of this plan, the IT will develop these measurement capabilities.

Table Components:

- Key initiatives — strategic objectives. While they may be single deliverables, more often than not, they will encompass a series of action items and associated critical success factors.
- Responsible parties — name of the individual responsible for the initiative, as well as others from within or outside of the team who play an important role in the delivery and success of that particular initiative.
- Timetables/resources — typically the delivery date and any special resource (i.e., cost) or revenue considerations.
- Performance metrics — actual measures or methods of measurement. The focus here is on results (outcome) measures and *not* activity measures. Performance metrics are often the same as the critical success factors for a product or service.
- Self-review — provided by the responsible party; an accounting of key measure results (e.g., if customer satisfaction is a key measure of a product's success, this column would include data conveying the change in customer satisfaction directly attributable to that product).
- Manager's review — comment by the appropriate executive or process owner.

APPENDIX C: PMO VALUE CALCULATION — MODEL AND TEMPLATE

Value/Cost Categories	Amount of Financial Benefit/Non-PMO IT Costs	IT Costs Avoidance Associated with Service Delivery Risk Mitigation	IT Costs Avoidance Associated with Project Delivery Risk Mitigation	PMO Investments in Services and Risk Mitigation	Outcomes (Net Value of Positive Outcomes and Risk Avoidance Less PMO Costs)	Comments
Planning Process						
Requirements						
Benchmarking						
Metrics/targets						
Template preparation						
Process management						
Presentations						
Documents/updates						
Reporting						
Service Level Management						
Customer Group A						
Requirements						
Metrics/targets						
Template preparation						
Process management						
Documents/updates						
Reporting						

Appendix C: PMO Value Calculation — Model and Template ■ 265

Customer Group B

Requirements
Metrics/targets
Template prep
Process management
Documents/updates
Reporting

Customer Group C

Requirements
Metrics/targets
Template prep
Process management
Documents/updates
Reporting

Customer Group D

Requirements
Metrics/targets
Template prep
Process management
Documents/updates
Reporting

Value/Cost Categories	Amount of Financial Benefit/Non-PMO IT Costs	IT Costs Avoidance Associated with Service Delivery Risk Mitigation	IT Costs Avoidance Associated with Project Delivery Risk Mitigation	PMO Investments in Services and Risk Mitigation	Outcomes (Net Value of Positive Outcomes and Risk Avoidance Less PMO Costs)	Comments
Customer Group E						
Requirements						
Metrics/targets						
Template preparation						
Process management						
Documents/updates						
Reporting						
Project Delivery Management						
Project A						
Strategic risk						
Financial risk						
PM risk						
Technology risk						
Change management risk						
Quality risk						

Project B

Strategic risk
Financial risk
PM risk
Technology risk
Change management risk
Quality risk

Project C

Strategic risk
Financial risk
PM risk
Technology risk
Change management risk
Quality risk

Project D

Strategic risk
Financial risk
PM risk
Technology risk
Change management risk
Quality risk

Value/Cost Categories	Amount of Financial Benefit/Non-PMO IT Costs	IT Costs Avoidance Associated with Service Delivery Risk Mitigation	IT Costs Avoidance Associated with Project Delivery Risk Mitigation	PMO Investments in Services and Risk Mitigation	Outcomes (Net Value of Positive Outcomes and Risk Avoidance Less PMO Costs)	Comments
Project E						
Strategic risk						
Financial risk						
PM risk						
Technology risk						
Change management risk						
Quality risk						
Project F						
Strategic risk						
Financial risk						
PM risk						
Technology risk						
Change management risk						
Quality risk						

Appendix C: PMO Value Calculation — Model and Template

Technology Platform Standardization and Rationalization

Hardware costs
Software costs
Support costs
Contractor costs
Customer servicing costs
Staff training costs
Other costs (facilities, back-up, etc.)

Staffing Costs

Management training
Technical training
Retention
Recruiting

APPENDIX D: SERVICE LEVEL AGREEMENT TEMPLATE

DRAFT SERVICE LEVEL AGREEMENT
(NAME OF BUSINESS UNIT)

[IT organization name], version 1.0

Business Unit Sponsor:
Business Unit Working Client(s):
IT Division, Customer Relationship Executive:
October 17, 2001
Version: 1.0

DRAFT SERVICE LEVEL AGREEMENT

TABLE OF CONTENTS

Introduction 273
Overview of the Division of Information Technology Services 274
Service Level Overview 274
    Business Application Systems or IT Services Subsumed
    within this SLA 275
    Exclusions 275
Business Unit Responsibilities 275
Maintenance (What IT Does on Behalf of the Customer) 276
    Definitions 276
Maintenance Service Level Definitions 277
Production Systems Problem Resolution 277
Support Service Levels 278
Problem Escalation 278
    Priority 1 278
    Priority 2 279
    Priority 3 279
Metrics 279
    Measurements and Reporting 280
    Reporting 280
    Communicating Service Level Performance 280
    Problem Escalation 280
Appendix 280
    SLA Process Role Definitions 280

INTRODUCTION

The purpose of this Service Level Agreement (SLA) is to establish agreed-upon service levels for the maintenance and enhancement of [name of business unit's] existing information technology assets, for the XXXX fiscal year. Information Technology's (IT) objective is to satisfy you, our customer, by providing dependable service and timely responses to your business priorities and information technology problems in the most cost effective manner.

The following Information Services associates are responsible for the successful delivery of IT products and services under this agreement.

Associate Responsible	Extension and E-mail Address	SLA Role and Title
Name, Title	phone x. e-mail	IT Customer Relationship Executive
Name, Title	phone x. e-mail	IT Business Operations
Name, Title	phone x. e-mail	IT PMO

Please note:

SLAs are living documents, amended on an as-needed basis to reflect changes based on your needs. Should you find it necessary to review or modify your XXXX SLA, either due to unusual circumstances or as additional requirements arise, please consult your customer relationship executive.*

---

* SLA activities encompass both nondiscretionary IT services, such as vendor-based software licensing, maintenance, and support, and the discretionary work associated with system and Web site enhancements. Each business unit's SLA resource allocation is set to meet all nondiscretionary service requirements. Discretionary work and hence resources are allocated based on past activity, anticipated needs, and explicit enterprise priorities. The total IT investment in these services is set by the enterprise's budgeting process in the late fall and winter of each year for the coming year.

OVERVIEW OF THE DIVISION OF INFORMATION TECHNOLOGY SERVICES

Name of unit
Name of managing executive
Phone number
E-mail address

Brief discussion of services provided and operating hours

Name of unit
Name of managing executive
Phone number
E-mail address

Brief discussion of services provided and operating hours

Name of unit
Name of managing executive
Phone number
E-mail address

Brief discussion of services provided and operating hours

etc.

SERVICE LEVEL OVERVIEW

This SLA addresses *all* the business-as-usual services provided to [name of business unit] by Information Services, encompassing the following elements:

- Identifies information systems covered by the agreement
- Establishes the level of service for system maintenance and support, including escalation procedures
- Provides criteria for measuring service level performance
- Identifies the process for communicating service level performance

Business Application Systems or IT Services Subsumed within this SLA

System or Service Type	IT Division Provider(s)	IT Division Service Provider(s)
• [e.g., Computer Help Line] Availability: [e.g., Monday – Thursday: 7–7; Friday: 7–6; Saturday: 9–4]	• (e.g., ITCS Call Center)	• [e.g., Lori Karas, Robert Rose]
•	•	•
•	•	•
•	•	•

Exclusions
This agreement excludes work defined as new development or project work in excess of $10,000 in cost. These latter efforts will be addressed through the IT synchronization and project management processes, conducted in close coordination with the enterprise's overall planning and prioritization processes (both long-term unit plans and annual business plans). These processes are documented elsewhere. Contact the project management office for details.

BUSINESS UNIT RESPONSIBILITIES

[name of business unit] is responsible for:

1. Operating within the information technology funding allocations and funding process as defined by the enterprise's governance, planning, and budgeting processes
2. Working in close collaboration with the designated IT customer relationship executive (CRE) to frame this SLA and to manage within its constraints once approved as part of the budget for the aforementioned fiscal year
3. Collaborating throughout the life cycle of the project/process to ensure the ongoing clarity and delivery of business value in the outcomes of the IT effort, including *direct participation in and ownership of* the quality assurance acceptance process
4. Reviewing, understanding, and contributing to systems documentation, including project plans and training materials, as well as any IT project/service team communications, such as release memos

5. Throughout the life cycle of the process, evaluating and ultimately authorizing business applications to go into production
6. Distributing pertinent information to all associates within the business unit who utilize the products and services addressed in this SLA.
7. Ensuring that business unit hardware and associated operating software meet or exceed the business unit's system-complex minimum hardware and software requirements
8. Reporting problems using the problem reporting procedure detailed in this SLA, including a clear description of the problem
9. Providing input on the quality and timeliness of service
10. Prioritizing work covered under this service agreement and providing any ongoing prioritization needed as additional business requirements arise
11. Employing enterprise information technology standards and architectures whenever possible and recognizing the total cost of ownership (TCO) implications of failing to observe these standards

MAINTENANCE (WHAT IT DOES ON BEHALF OF THE CUSTOMER)

Definitions

Maintenance is defined as any activity performed at the discretion of IT, which invests in and preserves the *value to the customer* of an existing application and environment including:

- Defect correction — correction of critical defects found in a deployed application that inhibit [name of business unit] from meeting its production system availability or performance requirements. Examples of defect correction activities include responding to production calls for batch systems running overnight or installing system bug fixes.
- Retooling — any change required to a [name of business unit] business application due to an upgrade of an infrastructure product. An example would be maintenance necessitated by changes in the production version of PeopleSoft, Lotus Notes, etc.
- Asset protection — business application upgrades in keeping with the vendor releases. An example would be the release of a new version of Microsoft Office, PeopleSoft, PowerFAIDS, etc.

- Disaster recovery procedures — supporting the business unit in developing its disaster recovery plan and participating in any disaster recovery testing.
- Required by external agencies — activities required by external and internal audit, federal agencies, etc. For example, the process to move and maintain applications into a formal change management environment.
- Applied research and feasibility analysis — as part of both SLA and project work for the business unit, Information Services will conduct assessments of IT products, services, and processes to determine their appropriateness in line with business unit needs.
- Infrastructure and related production support — work associated with the implementation of business system applications and end-user desktops.
- System support — Tier 2 and 3 support associated with both the break/fix of existing business application systems and the problems encountered by end-user customers in using these business application systems.
- Information security — identification, assessment, management, and remediation of threats, vulnerabilities, and risks to the electronic information assets of the institution.

MAINTENANCE SERVICE LEVEL DEFINITIONS

- When problems occur, the following Problem Escalation section will govern IT team responses. This work will take priority over project and other support work until the problem is resolved.
- Retooling and platform upgrades will be scheduled as appropriate to meet business schedules and requirements. As a common practice, such upgrades would typically reach production through regularly scheduled releases of bundled upgrades and enhancements.

PRODUCTION SYSTEMS PROBLEM RESOLUTION

Systems staff is available for production problem resolution 24 hours a day, 7 days a week, on an on-call basis. For problem resolution, contact the Computer Help Line (xxx-xxx-HELP).

SUPPORT SERVICE LEVELS

(Tier 2 and 3 support and problem resolution)

Support Service Levels and Response Time According to Level of Severity

Severity[1]	Customer Impact	Customer Response	Resolution
Code 0 — Classroom response	Information technology, telecommunications, or multimedia failures that immediately impact classroom delivery	Immediate[2]	ASAP
Code 1 — Catastrophic	Global campus service halted Notify: Customer Services director within 15 minutes	Within 30 minutes	ASAP
Code 2 — Urgent	Subset of campus impacted Notify: Customer Services manager or ResNet manager within 15 minutes	Within 60 minutes	ASAP
Code 3 — Important	Individual impacted	Within 4 hours	24 hours
Code 4 — Non-critical	Scheduled/planned work Informational only	72 hours' notice at a minimum	As scheduled
Code 5 — Other	Maintenance of IT systems and tools	N/A	As time allows

[1.] Severity is as determined by the computer help line/ITCS and the customer.
[2.] At the present time, Campus Media Services fields all classroom calls and either dispatches one of its own personnel or works with the ITCS to respond to faculty needs.

Note: For action leading to problem resolution to occur, the problem must be called in to the Computer Help Line (xxx-xxx-HELP).

PROBLEM ESCALATION

Priority 1

- *Definition* — application is unavailable to anyone at a site (e.g., Boston, New York, Chicago).
- *Response time* — work will begin immediately and continue until resolved.

- *Responsibilities*:
  - IT customer relationship executive (CRE) — resolves problem and communicates to all who are affected at least daily until resolved.
  - Working client — works alongside CRE until the matter is resolved.
  - Partner providers — other IT teams and third parties will provide technical assistance as appropriate.

Priority 2

- *Definition* — application is unavailable for a group of users within a site.
- *Response time* — a response will be provided within one business day. A recommended solution will be provided within three business days if there are *no* outstanding priority 1s. Finding a solution to a priority 2 problem will not begin until all priority 1 problems that impact the priority 2 issue's resolution have been resolved.
- *Responsibilities*:
  - IT customer relationship executive — sends acknowledgement of problem. Resolves problem and communicates status to all who are affected.
  - Working client — works alongside CRE until the matter is resolved.
  - Partner providers — other IT teams and third parties will provide technical assistance as appropriate.

Priority 3

- *Definition* — application generates appropriate results but does not operate optimally.
- *Response time* — improvements addressed as part of the next scheduled release.
- *Responsibilities*:
  - IT customer relationship executive — communicates needed changes.
  - Other process participants — as part of the regular system upgrade cycle.

METRICS

IT collects and reports on metrics for two reasons:

- To measure and demonstrate proper IT resource management in line with enterprise priorities

- To measure and demonstrate the provision of service levels in line with customer requirements

Measurements and Reporting

The IT customer relationship executive (CRE) for [name of business unit] will work with all appropriate IT service delivery personnel to measure and report on IT performance.

Reporting

- System maintenance and support, including production modifications, system maintenance, bugs fixed, and enhancements delivered — monthly SLA review
- Service level satisfaction — monthly discussion between the sponsor/working client(s) and the IT CRE and as measured through the IT operations report process
- Status of customer-related IT projects — monthly review of scorecards

Communicating Service Level Performance

Regular status review meetings, no less than monthly, based on the above reporting. The CRE will involve his or her own staff and partner providers as required while relying on his or her working clients to involve end users from the customer community as appropriate. The key information gathering and reporting tools in this process will include the monthly IT operations report, IT customer satisfaction surveys, and IT project scorecards. When issues or action items arise during status meetings, the CRE is charged with following up on any open items or issues raised.

Problem Escalation

See page 274 for details.

APPENDIX

SLA Process Role Definitions

- Sponsor — the executive leader of the business unit who ultimately approves the funding for the business unit's IT work.
- Working clients — those business-unit managers who work alongside their Information Services counterparts to define, develop, and deliver IT products and services to the business unit.

- IT customer relationship executive (CRE) — the IT executive ultimately responsible for the satisfactory delivery of the IT commitments consolidated under this SLA.
- IT business enterprise operations — IT's project/service management center of excellence will review and approve all SLAs from a best-practices perspective to ensure that they align with the highest standards of service delivery balanced against available IT and enterprise resources.
- IT business operations — IT's business office will review and approve all SLAs from a financial perspective to ensure that they align with overall commitments to the enterprise community.

APPENDIX E: PROJECT MANAGEMENT LIFE-CYCLE FRAMEWORK

PROJECT PHASE	PHASE DESCRIPTION	PHASE TASKS	DEVELOPMENT PHASE DELIVERABLES	PROCIT PERSONNEL RESPONSIBILITIES
Commitment	Project planning, initiation, & commitment	A. Receive authorized request from executive sponsor and working clients B. Analyze project/service request C. Define project scope D. Determine approach E. Define project team assignments F. Model resource requirements G. Hold commitment review, obtain signoffs H. Obtain resource commitments I. Create the project control process J. Create project scorecard K. Hold kickoff meeting	L. Commitment document with signoffs by executive sponsor, working clients, and all IT organization partner providers for the project M. Project plan N. Project budget O. Master project schedule entry P. Scorecard (monthly updates) Q. Scope change control forms R. Deliverable acceptance forms	S. Project director and manager meet iteratively with project sponsor and working client(s) to define scope and resources T. Project director, project manager, and working clients develop commitment document and draft project plan U. Project director or project manager obtain input on commitment document and plan from IT organization partner providers

Appendix E: Project Management Life-Cycle Framework ■ 285

Analysis, Part 1 Business requirements analysis

A. Confirm project scope
B. Establish customer needs
C. Assess current business processes
D. Define quality assurance (QA) evaluation process and scope
E. Identify gaps and opportunities for IT-enabled process changes
F. Develop launch strategy and preliminary implementation plan
G. Model process and data requirements
H. Develop project measures and metrics
I. Review and sign off on business requirements
J. Change management process in place

K. Business requirements document with business benefits/ROI analysis if appropriate
L. Functional requirements document for IT-enabled business solution
M. Process model/use cases
N. Page schematics for Web solutions
O. Review scope changes
P. Updated commitment document
Q. Deliverable acceptance form signed by the customer at each project phase

R. Project team, led by project director, with management and analysis services provided by PMO (if applicable); technical analysis provided by partner providers and external resources as needed
S. Project team, including working clients, to develop customer service strategy and plan, including help desk training and customer feedback plan
T. Project team ensures deliverables acceptance by the customer and formally ends the phase
U. Project team ensures that QA reviews phase outcomes for the subsequent development of test plans and scripts

PROJECT PHASE	PHASE DESCRIPTION	PHASE TASKS	DEVELOPMENT PHASE DELIVERABLES	PROCIT PERSONNEL RESPONSIBILITIES
Analysis, Part 2	Technical and infrastructure requirements analysis	A. Analyze solution alternatives B. Recommend technical solution C. Architect and model initial IT solution D. Define QA evaluation process and scope E. Validate against organization's existing IT architectural standards and best practices F. Gather technical requirements and discuss implications with appropriate IT infrastructure management personnel (i.e., network, server, storage, production services impacts) G. Review and sign off on technical specification H. Review and approve recommended solution with working clients and IT partner providers	I. Complete technical requirements questionnaire form with stakeholder input (both working clients and partner providers as required) J. Technical specification document K. Updated commitment document L. Deliverable acceptance form signed by the customer at each project phase	M. Project team to engage internal and external partner providers in process N. QA, partner providers, external parties, et al., to provide technical input O. In particular, the project team must involve IT infrastructure personnel (e.g., system services and network services) at this stage of the project and thereafter as changes occur in the technical architecture and deployment plans for the project P. Project team ensures deliverables acceptance by the customer and formally ends the phase Q. Project team ensures that QA reviews phase outcomes for the subsequent development of test plans and scripts

Appendix E: Project Management Life-Cycle Framework ■ 287

Design, Part 1 | Solution definition and design

A. Establish business process and technical frameworks
B. Specify systems/subsystems involved in the delivery of the IT solution
C. Specify data requirements
D. Design databases
E. Specify/design data conversion process
F. Complete documentation of business process changes and system specifications
G. Specify operational requirements, including training for IT and customer support units and end users
H. Define QA testing and evaluation
I. Define test strategy
J. Define development infrastructure
K. Training and support needs defined
L. Detailed functional specifications
M. Preliminary visual designs and screenshots
N. Development infrastructure specifications
O. QA/test scripts and infrastructure specifications
P. Workflow design and organization plan
Q. Adjustments to technical specification document
R. Deliverable acceptance form signed by the customer at each project phase
S. Capture any scope changes in commitment document
T. Draft service level agreement (SLA) matrix and associated IT operational service delivery agreement (OSDA) metrics
U. Project manager to ensure review and signoff on detailed functional specification
V. Project manager or business analyst to carry out usability planning with working clients, other customers, and the project team
W. Release strategy developed by QA working with project manager
X. Review and approve final specifications and agree on change management process by project team and working clients
Y. PMO personnel to modify technical specifications as required
Z. Project director ensures deliverables acceptance by the customer and formally ends the phase

PROJECT PHASE	PHASE DESCRIPTION[1]	PHASE TASKS	DEVELOPMENT PHASE DELIVERABLES	PROCIT PERSONNEL RESPONSIBILITIES
Design, Part 2	Prototype[1]	A. Create visual prototype based on business and technical requirements; employ "sandbox" IT environments as appropriate B. Conduct focus groups C. Adjust solution design as necessary, based on feedback D. Place orders for IT hardware and software as appropriate	E. Nonworking prototype an aid to understanding requirements F. Launch plan G. Focus groups findings document H. Purchase orders I. Deliverable acceptance form signed by the customer at each project phase J. Capture any scope changes in commitment document and project plan	K. Working clients review and provide feedback L. Working clients review and sign off on prototype and also visual designs/prototypes M. Working clients review and approve updated project plan and commitment document N. Launch planning with customers and project team, including IT infrastructure personnel O. Project director ensures deliverables acceptance by the customer and formally ends the phase

[1]. May not apply in all cases.

Appendix E: Project Management Life-Cycle Framework ■ 289

Infrastructure | Infrastructure requirements for development, testing, and production

A. Review project technical specifications with IT infrastructure and QA personnel
B. Given the technical specifications, identify development environment
C. Secure development lab resources
D. As development proceeds, anticipate implications of changes in technical requirements that impact IT test and production environments
E. Communicate implications to QA regarding test lab
F. Communicate implications to IT infrastructure personnel regarding production
G. Take steps during development to ensure that both test and production platforms are properly resourced to receive the project's system at the times stated in the project plan
H. Architectural review
I. Business continuity and disaster recovery review
J. Development lab specifications
K. Test lab specifications (or at least the beginnings of gap analysis between current capabilities and projected needs once the project proceeds to QA)
L. Production environment specifications (or at least the beginnings of gap analysis between current capabilities and projected needs once the project proceeds to production)
M. Purchase orders as appropriate
N. Synchronization of test lab and production change scheduling in line with the project plan
O. Delivery plan adjustments as appropriate
P. Deliverable acceptance form signed by the customer at each project phase
Q. Capture any scope changes in commitment document and on project plan
R. Defined backup, restore, and disaster recovery standards/thresholds
S. Review and signoff by the development lab coordinator
T. Review and signoff by IT infrastructure personnel
U. Review and signoff by QA
V. Review and signoff by the IT team that will host the application once it is in production
W. IT executive approval of all software acquisitions
X. IT executive approval of all hardware acquisitions
Y. Architectural review and signoff by IT management
Z. IT business officer approval of purchase orders

PROJECT PHASE	PHASE DESCRIPTION	PHASE TASKS	DEVELOPMENT PHASE DELIVERABLES	PROCIT PERSONNEL RESPONSIBILITIES
Development	Development	A. System(s) installation and modification B. Technical development of database, middleware, and front-end components C. Data feeds/cleanup D. Develop reporting services and reports E. Unit testing F. Develop test planning and test script G. Support staff training and document development H. Prepare customer service and marketing plans I. Reengineer business processes J. Screenshots documented K. Support staff and end-user training developed L. Communication and marketing strategies developed	M. Application that conforms to functional specifications N. User documentation O. Technical documentation P. Reporting package (for canned reports) and reporting/query tools for end-user *ad hoc* reporting Q. Business workflow procedures document R. User training package S. Deliverable acceptance form signed by the customer at each project phase T. Capture any scope changes in commitment document U. Technical design document V. Training program descriptions	W. Project team develops solution in partnership with internal and external partner providers X. Data lead ensures all data feeds in place and that the data is clean/normalized for use by application Y. PMO personnel and working clients provide business process and application documentation Z. PMO personnel provide business cases for test plan AA. Project team to review and approve reporting package with working clients AB. PMO personnel to coordinate technical support and user training and customer service and marketing plans; ensure partner provider agreement AC. Project director to provide metrics for stress and acceptance testing

Appendix E: Project Management Life-Cycle Framework ■ 291

Certification	QA testing and release management	A. Perform functional QA tests B. Perform technical tests C. Perform usability testing D. Perform system integration testing E. Perform stress testing F. Perform acceptance testing G. Fix bugs! H. Initiate communications and marketing strategies I. Collect information about future enhancements as part of the development of a release strategy for the new or revised IT product or service	J. Testing approaches/scripts K. Implemented system in test environment L. QA certification M. Final implementation plan N. Support team training validation O. Customer training validation P. Refined application per functional QA, technical, usability, system integration, environmental, and acceptance testing activities Q. Deliverable acceptance form signed by the customer at each project phase R. Capture any scope changes in commitment document	S. QA to review and approve test acceptance criteria T. QA to review and approve test approach and plan U. QA to perform functional, technical, usability, system, and stress testing, typically within QA lab V. Perform customer acceptance testing, typically within customer environments W. QA certifies ready for production X. Project manager to communicate customer feedback to project team and working clients Y. Project director to communicate with users regarding pilot/full launch Z. Executive sponsor and working clients sign off on system AA. Risk review held with IT management for formal "Go/No Go" signoff

PROJECT PHASE	PHASE DESCRIPTION	PHASE TASKS	DEVELOPMENT PHASE DELIVERABLES	PROCIT PERSONNEL RESPONSIBILITIES
Launch	Pilot[1]	A. Extend the implementation of communications and marketing strategies B. Implement pilot system or system (as appropriate) C. Regression testing completed and all bug fixes either addressed or plans in place for resolution D. Transition to servicing team E. For Web sites, load Web assets and site content (ongoing process) F. Training underway; user support in place	G. Implemented production-ready system H. Launch-ready organizational packages I. SLAs, including future enhancements, ongoing maintenance, release planning, and support costs J. Augmented future enhancements document K. User training and documentation in place	L. Project team monitors pilot progress M. Revise launch plans with working clients and project team N. Project team reviews pilot feedback; determines need to incorporate feedback prior to launch O. Project team risk review held with management for formal "Go/No Go" signoff P. IT customer services to publish training schedules and documentation Q. PMO personnel to adjust IT OSDAs and customer SLAs.

[1]. This step does not apply in all project instances.

Appendix E: Project Management Life-Cycle Framework ■ 293

Release	Go live	A. Go live signoff by QA B. Implement production system C. For Web sites, load Web assets and site content (ongoing process) D. As part of hand-off to ongoing IT service provider, assist in development of release strategy that defines future enhancements as part of a 1–2 year view of the product/service offering E. Meet with your team to bring closure	F. Documentation completed G. Communication/press release H. Code archived I. All technical, systems, and user documentation archived J. All project artifacts in IT knowledge repository K. Project signoff by customer L. Release strategy	M. Project team to assist with user training and launch activities N. Customer signoff on deliverables O. Appropriate IT manager to meet with project team and discuss lessons learned P. Project director to monitor ongoing feedback mechanisms — e-mail, calls to service center, metrics, and measures — and, as appropriate, initiate steps to enhance program and services Q. Project manager to deposit all major project documents in an IT project knowledge management repository I. Project director ensures that all changes are well communicated in advance to customers and IT personnel
Sunset	Retire existing and replaced platforms	A. Recovery planning for legacy systems B. Data preservation for historical reporting/audit purposes C. Uninstall software as it comes offline D. Uninstall hardware as it comes offline	E. Postimplementation celebration F. Update IT organization asset inventories G. Adjust OSDAs and SLAs accordingly H. Lessons-learned document in IT knowledge repository	

APPENDIX F: PROJECT LEADERSHIP QUESTIONNAIRE FOR CHANGE MANAGEMENT PROJECTS

Note: This document was developed by the Babson College Management Team in October 1993, from a checklist originally prepared by Allan R. Cohen and Rosabeth M. Kanter; the tool was subsequently updated by Richard M. Kesner.

THE LEADERSHIP COMPONENT

- Who is leading the project? _____
- Is that person the most appropriate responsible party to ensure success? _____
- Does the leader genuinely believe in the project and want it to succeed? _____
- Does the leader have the necessary skills for success?
  - Demonstrates genuine respect for people throughout the organization _____
  - Listens well _____
  - Demonstrates political savvy _____
  - Possesses connections with others critical to the project's success _____
  - Demonstrates salesmanship _____
  - Inspires trust _____
  - Communicates readily and well _____
  - Demonstrates patience _____
  - Focuses on goals and their realization, even in chaotic process _____
  - Demonstrates flexibility _____
  - Follows through on assignments/attentive to detail _____
  - Possesses high energy _____
  - Is willing to delegate _____
  - Is willing to share credit for project deliverables _____
- Has project leadership been legitimized, as appropriate, with
  - Executive management? _____
  - Board(s) of governance? _____
  - Standing committees and task forces? _____
  - Project leader's boss? _____
  - Project leader's peers? _____
  - Project participants? _____
- If the project is to be executed by a task force, have they identified/legitimized a leader? _____
- Is leadership accessible and open to input from throughout the organization? _____
- Are there opportunities for others to take on leadership responsibilities when they are ready and as the opportunity presents itself? _____

ORGANIZATIONAL LINKAGES

- Has a compelling case for change been made? ____
- Has the case for change taken into consideration other changes already underway around the organization? ____
- Is there a formal steering committee or informal advisory committee of key senior managers/stakeholders? ____
- Are there clear procedures for obtaining their opinions and keeping them informed? ____
- Has the project team completed a stakeholder analysis, measuring the relative impact on stakeholders of the following variables:
  - Status? ____
  - Clout/power/influence? ____
  - Rewards? ____
  - Opportunities/challenges? ____
  - Visibility? ____
  - Self-image? ____
- Is there a working group/task force/change committee involving stakeholders? ____
- Is there a clear and convincing linkage between the proposed project and the enterprise's strategy, goals, and objectives? ____
- Does this discussion include a consideration of the needs and concerns of long-standing members of the enterprise who may view proposed changes as a critique of past performance and contributions? ____
- Is there a process for communicating progress/developments to stakeholders? ____
- Is there a process for communicating progress/developments to non-stakeholders? ____
- Have the supervisors of change process participants
  - Agreed to the assignment? ____
  - Appropriately adjusted goals and objectives? ____
- Have provisions been made for participants to give time to the project without having
  - Their other work suffer? ____
  - Their co-workers suffer? ____
- Are the project's goals and objectives consistent with the goals and objectives of the enterprise? ____
- Have other areas of the enterprise that might affect or be affected by the project been
  - Identified? ____
  - Informed? ____
  - Incorporated into the project design when possible? ____

- Are there mechanisms in place to ensure the general dissemination of project goals, activities, and results? ____
- Are the linkages in place between the project steering committee and associated task forces, etc.? ____
- Do steering committee and task force members understand their individual responsibilities to keep their colleagues/constituencies informed? ____
- Do steering committee and task force members understand their individual responsibilities to obtain input from their colleagues/constituencies? ____
- Is the approval process for decisions and action/implementation clearly articulated and consistently communicated to stakeholders? ____
- Has the enterprise committed to a realignment of resources in line with project priorities and objectives? ____

EXPECTATIONS

- Has the project team diagnosed employee attitudes toward the proposed change? ____
- Will they participate in the change process? ____
- Has the team determined the optimal time for project kickoff? ____
- Has the team determined the optimal pace for the changes envisioned in the project? ____
- Has a clear message gone out to the enterprise concerning realistic project accomplishments? ____
- Has sufficient information about potential payoffs been disseminated so that employee expectations are high enough to ensure general involvement? ____
- Have likely difficulties and a full timetable been communicated so that expectations are not unrealistically high? ____
- Have stakeholders had an opportunity to voice their concerns? ____
- Are project participants at all levels clear about the extent of their
  - Mandate? ____
  - Authority? ____
  - Budget? ____
  - Nonfinancial resources? ____
  - External/internal partners (teammates)? ____
  - Time commitment? ____
- Have sufficient resources been allocated to complete the project in a timely and effective manner? ____

PROJECT PARTICIPATION

- Is there enough organizational flexibility to allow for participation by participants? ____
- Is participation at all levels voluntary? ____
- Are there mechanisms in place for informing all those eligible to volunteer? ____
- Is the selection process for participation fair? ____
- Is it perceived to be fair? ____
- Are at least some participants of high status relative to their peers to ensure that participation is viewed as a reward? ____
- Is there a way for participants' supervisors to stay informed and make suggestions?
- Are the skills required of participants to participate effectively
  - Known? ____
  - Known by them? ____
  - Learnable through established/prearranged means? ____

REWARDS

- Have arrangements been made to reward participants for their
  - Time? ____
  - Ideas? ____
  - Commitment? ____
- Has the project's impact on existing reward systems been analyzed? ____
- Has the reward of participating
  - Been recognized? ____
  - Been taken into account for its impact on nonparticipants? ____
- Can the project succeed if the formal award system remains unchanged? ____
- Are changes needed in the existing reward system to support project goals? ____
- Have required changes in the existing reward system been
  - Explored? ____
  - Made? ____

APPENDIX G: IT PROJECT RISK MANAGEMENT MATRIX

Risk and Feasibility Prediction Tool v2.0

Risks:	Weighting	Score	Weighted Score	Comments
Strategic Risk				
o Market image impact of late/non delivery				scaled score
o Competitive impact of late/no delivery				scaled score
o Sales/revenue (ROO) impact of late/non delivery				scaled score
o Business value of project agreed to by stakeholders				yes/no
Requirements Risk				
o Requirements clearly defined and agreed to				yes/no
o Project results traceable to requirements				yes/no
o Project requirements stable				yes/no
Technology Risk				
o Maturity/availability of technology				scaled score
o Complexity of system integration				scaled score
o Compliance with industry standards				scaled score
o Compliance with organization's I/T architecture and standards				scaled score
o Difficulty of deployment				scaled score
o Difficulty of ongoing support				scaled score
o Performance implications on related applications				scaled score
o Security issues				yes/no
o Data integrity issues				yes/no
Vendor/Partner Risk				
o Stability/reliability of vendor/partner				scaled score
o Duration of relationships				scaled score

Implementation Risk				
o Track record on past projects				scaled score
o Benchmark of degree of difficulty				scaled score
o Life-cycle methodology in place				yes/no
o Change management review process in place and followed				yes/no
o Stakeholder commitment				yes/no
o Availability of skilled resources				scaled score
o Project control mechanism(s) in place				yes/no
o Project performance metrics in place				yes/no
o Impact on other I/T systems				yes/no
o Performance metrics in place/use				yes/no
Adoption Risk				
o Organization's willingness to adopt				scaled score
o Customers' willingness to adopt				scaled score
o Effective communication process in place among stakeholders				yes/no
o Reviews with stakeholders are part of standard operating procedures				yes/no
Financial Risk				
o Soundness of budget				scaled score
o Security of funding				scaled score
o Accurate/timely status reporting				yes/no
o Actuals to budget gap				percentage gap/scaled score

APPENDIX H: COMMITMENT DOCUMENT TEMPLATE

[PROJECT NAME] COMMITMENT TEMPLATE

The Project Management Office
[commitment author]
[date]
[Version # 1.0]

[COMMITMENT TITLE]

TABLE OF CONTENTS

REQUEST 307
   1. Project Name: [commitment name] 307
   2. Commitment Type 307
   3. Business Priority Assignment 307
   4. Brief Statement of Business Problem or Opportunity 307
SPECULATION 308
   5. Business Value Statements 308
   6. Commitment Tie to Strategic Business Initiative 308
   7. Project Business Definition 308
   8. Critical Success Factors — Conditions of Customer Satisfaction .. 309
   9. Impact Statement 309
   10. Sunset Statement 309
   11. Assumptions, Constraints, and Open Issues 309
   12. Risk 310
   13. Statement of Position 98-1 310
   14. Preliminary Estimate of Cost, Duration, and Effort 310
   15. Roles and Responsibilities 311
   16. Partner Provider Signature 312
OFFER 312
   17. Offer Expiration Date 312
   18. Commitment Offer Review 312
COMMITMENT 313
   19. Signature Approval or Disapproval 313
   20. Revision History 314
   21. Signature Approval of Change 314
   22. Customer Satisfaction Questionnaire 315

REQUEST

1. Project Name: [commitment name]

Commitment Level: _ Request _ Speculation _ Offer _ Commit _ Declined _ Complete

Business unit:

Project's executive sponsor (business unit head):

Working clients:

IT project director:

Project management office representative(s):

Products/services/systems/application name(s):

2. Commitment Type

_ Proprietary _ Commercial _ Hybrid _ Platform

3. Business Priority Assignment

_ Enterprise	_ Line of Business/Knowledge Center
_ **High** (Critical for [Enterprise]'s business continuation)	_ **High** (Competitive opportunity; legal/industry requirement; provides significant economic or customer satisfaction benefits)
_ **Medium** (Broad-based economic benefit to [Enterprise])	_ **Medium** (High payback potential; major contributor to customer satisfaction goals)
_ **Low** (Benefit to [Enterprise], or is a specific, mandated legal requirement)	_ **Low** (Key contributor to customer satisfaction goals)

4. Brief Statement of Business Problem or Opportunity

SPECULATION

5. Business Value Statements

Business Improvement	Major	Minor	None	Business Value Statement (In Support of the Improvement)
1. Increase revenue				
2. Decrease cost				
3. Avoid cost				
4. Increase productivity				
5. Improve time to market				
6. Improve customer service/value				
7. Provide competitive advantage				
8. Reduce risk				
9. Improve quality				
10. Other (describe)				

6. Commitment Tie to Strategic Business Initiative

7. Project Business Definition

Delivery

Essential (must be delivered):

　1.
　2.

Desirable (delivery will be accepted without this functionality):

　1.
　2.

Optional (would be nice to have if it can be included within projected time and cost):

　1.
　2.

Exclusions:

1.
2.

8. Critical Success Factors — Conditions of Customer Satisfaction

Scope:

Quality:

Cost:

Time:

9. Impact Statement

Indicate *all* other applications that require changes as a result of this Commitment. List any other Commitments this project is dependent upon for successful completion.

10. Sunset Statement

List the application(s) to be sunsetted as a result of this Commitment. Estimate life span of this application.

1. Sunsets:
2. Life span of this Commitment's deliverables:

11. Assumptions, Constraints, and Open Issues

Assumptions:

1. This project will adhere to established information technology and application architectures (if this is not true, see section 12 — Risk) and will strive toward open, standards-based IT solutions.
2. The PMO project engineering framework will be used to build the project plan.
3. The PMO's standard project resource and time management tools will be employed by the project team to manage this project, to assign and account for resource utilization, and to report on project deliverables.
4.

Constraints:

1.
2.

Open Issues:

1.
2.

12. Risk

_ Breakthrough _ Platform _ Derivative _ R&D _ Support

Risk Management Matrix (Updates to this continue throughout life of Commitment.)

Potential Risk	Description of Risk	Resolution
Technology		
Financial		
Security		
Data integrity		
Business continuity		
Regulatory		
Business requirements		
Operational readiness		
Other (explain)		

Will this Commitment introduce any new technology that is not currently part of Information Technology architecture? (An answer of "yes" requires review and approval by the CIO.)

Approved by CIO:_____ date: _____

13. Statement of Position 98-1

(State whether the cost of this project will be expensed or capitalized.)

_ Expensed in current fiscal year _ Capitalized over [state how many] years

14. Preliminary Estimate of Cost, Duration, and Effort

Preliminary Estimated Cost:
  Best Case: _____ Planned: _____ Contingency: _____

Preliminary Estimated Duration (lapsed time in weeks or months):
  Best Case: _____ Planned: _____ Contingency: _____

Preliminary Estimated Effort (total project plan resource hours required — associates and contractors):
  Best Case: _____ Planned: _____ Contingency: _____

Confidence Level: __ Low   __ Medium   __ High

Maintenance Estimate (per year):

15. Roles and Responsibilities

(List the roles and responsibilities for this Commitment.)

Role	Name of Associate	Responsibility
NEU Team:		
Executive sponsor		Provides funding and is ultimately responsible for overall project delivery
Working client(s)		Works closely with IT team to ensure the proper synchronization among end-user requirements, project plans, and IT delivery
Project director		Oversees IT delivery and ensures working client satisfaction
Project manager(s)		Ensures adherence to best practices in the overall administration and delivery of the project(s)
Business analyst (EO)		Coordinates IT delivery and ensures both internal synchronization of IT resources and the synchronization of IT delivery with physical plant requirements
Project analyst (EO)		Assists in project delivery documentation, measurement, and reporting
Application lead		
Systems lead		
Data management lead		
Infrastructure lead		
Customer services lead		

Role	Name of Associate	Responsibility
Internal and External Partners:		
Vendor-based project management support		
Technical architect(s)		
Business process architect(s)		
Creative development/UI		
Development		
Training/ documentation		
QA/testing		
Infrastructure		
Security		
Other partner provider(s) — Hardware/Software:		

16. Partner Provider Signature

(Instructions: Hold a review of sections 1–12 with your partner providers and obtain their agreement to participate. Each partner provider's signature represents his or her agreement to participate in this effort.)

Partner Providers — Agreement to Participate

Organization/Representative	Signature	Date
		/ /
		/ /
		/ /
		/ /
		/ /
		/ /

OFFER

17. Offer Expiration Date

Offer expiration date:

18. Commitment Offer Review

Review Participants — Signature Approval

Review Participant	Signature	Date
Information Technology's Customer Relationship Executive: [name of CRE]		/ /
IT Business Operations: [name of business officer]		/ /
PMO: [name of PMO director]		/ /

COMMITMENT

19. Signature Approval or Disapproval

Signature Approval of Commitment

Organization/Representative	Signature	Date
Business Unit Sponsor:		/ /
Working Client (Business Rep):		/ /
Working Client (Business Rep):		/ /
Working Client (Business Rep):		/ /
Working Client (Business Rep):		/ /

Signature if Business Unit Declines to Proceed at This Time.

Organization/Representative	Signature	Date
Business Unit Sponsor:		/ /
Working Client (Business Rep):		/ /
Working Client (Business Rep):		/ /
Working Client (Business Rep):		/ /
PMO, Project Manager:		/ /
PMO, Director:		/ /

Explanation:

20. Revision History

(Instructions: *Any* changes to the information in this document *must* be itemized below. To validate the change, signature approval *must* be obtained.)

Description of Change:

21. Signature Approval of Change

Signature Approval

Organization/Representative	*Signature*	*Date*
Working Client (Business Rep):		/ /
PMO, Project Manager:		/ /
PMO, Director:		/ /
IT Business Officer:		/ /

Partner Providers — Approval of Change

Organization/Representative	*Signature*	*Date*
		/ /
		/ /
		/ /

22. Customer Satisfaction Questionnaire

Name: _____

	Very Satisfied	Satisfied	Neither Satisfied nor Dissatisfied	Dissatisfied	Very Dissatisfied	Not Applicable
How satisfied are you with:						
1. Critical Success Factors — Conditions of Satisfaction						
a. Scope: All agreed-upon business requirements are included.	5	4	3	2	1	0
b. Quality: The delivered product performs as expected.	5	4	3	2	1	0
c. Cost: The product was delivered within the agreed-upon cost.	5	4	3	2	1	0
d. Time: The product was delivered on the agreed-upon date.	5	4	3	2	1	0
2. Overall Satisfaction						
a. What is your overall satisfaction with the system/service?	5	4	3	2	1	0
b. What is your overall satisfaction with service provided by IT?	5	4	3	2	1	0

OPTIONAL — Completion of the following is optional; however, your responses will help us in our quest for continuous product improvement. Thank you!

3. System Data						
a. Data accuracy	5	4	3	2	1	0
b. Data completeness	5	4	3	2	1	0
c. Availability of current data	5	4	3	2	1	0
d. Availability of historical data	5	4	3	2	1	0
4. System Processing						
a. Accuracy of calculations	5	4	3	2	1	0
b. Completeness of functionality	5	4	3	2	1	0
c. System reliability	5	4	3	2	1	0
d. Security controls	5	4	3	2	1	0

How satisfied are you with:	Very Satisfied	Satisfied	Neither Satisfied nor Dissatisfied	Dissatisfied	Very Dissatisfied	Not Applicable
5. System Use						
a. Ease of use	5	4	3	2	1	0
b. Screen design	5	4	3	2	1	0
c. Report design	5	4	3	2	1	0
d. Screen edits (validity and consistency checks)	5	4	3	2	1	0
e. Response time	5	4	3	2	1	0
f. Timeliness of reports	5	4	3	2	1	0
6. System Documentation & Training						
a. Accuracy of documentation	5	4	3	2	1	0
b. Completeness of documentation	5	4	3	2	1	0
c. Usefulness of documentation	5	4	3	2	1	0
d. Training	5	4	3	2	1	0
e. Online help	5	4	3	2	1	0

Additional comments or suggestions? (Please use back of form.)

APPENDIX I: MASTER PROJECT SCHEDULE TEMPLATE

318 ■ The Hands-On Project Office

KEY:
Green=project on schedule; Yellow=project behind schedule but under control; Red=project in trouble; Blue=scorecard attached
White (C) = completed | no = no card | Purple = project on hold or pending
P = priority; 1 = highest priority; 2 = important but secondary IS priority; 3 = to be accomplished as IT resources allow

#	Project Name:	SC	P	2001 Start	J	F	M	A	M	J	J	A	S	O	N	D	J	F	M	A	M	J	J	A	S	O	N	D	J	F	M	A	M	J	J	A	S	O	N	D	Project Issues, Dependencies & Comments	Staffing Issues, Dependencies & Comments
				FY2001					FY 2002											FY 2003											FY 2004											
								2002												2003												2004										

Projects
Recently
Completed &
Removed:

Appendix I: Master Project Schedule Template

	Total # of Projects Listed =	Total # of Priority One Projects Listed =	Total # of Priority Two Projects Listed =	Total # of Priority Three Projects Listed =	Total # of Projects on hold or pending (purple) =	Total # of Projects on schedule (green) =	Total # of Projects behind schedule, but under control (yellow) =	Total # of Projects in trouble (red) =	Total # of Projects completed during past month =

APPENDIX J: GLOSSARY

Action (annual) plan a formal IT organization management process and document that captures the goals, objectives, responsible parties, and performance metrics for the unit's service and project delivery over the course of a given fiscal year and aligns these activities with the goals and objectives of the parent enterprise.

Alignment a management mindset and process whereby the activities and resource allocations of the IT organization map directly to the goals, objectives, and values of that body's parent organization.

Architected solution an IT system built upon the foundation of the enterprise's existing technology standards and architecture. Such solutions take full advantage of the technologies, business processes, and technical expertise already in place across the enterprise.

Architecture a set of standards, guidelines, and statements of direction that constrain the design of IT solutions for the purpose of eventual integration.

Artifacts (project) materials generated as a result of IT service and project management delivery efforts that document the process and its outcomes, such as project plans, budgets, commitment documents, specifications, scorecards, interview notes, and so forth.

Business analyst typically an IT staffer whose expertise lies in the documentation of customer business and functional requirements and the integration of this knowledge into the design and development of IT solutions.

Business process engineering those analytical and design activities concerned with the characterization of existing work processes within the enterprise and the restructuring of those workflows for efficiency, effectiveness, and to better leverage associated, enabling information technologies.

Call center (also, help desk) within an IT organization, the call center or help desk serves as the first point of contact between a customer requiring support with a problem or service requests. Typically, this first tier of customer support will address anywhere from 60 to 80 percent of the issues without recourse to Tier 2 or Tier 3 technical teams.

Certification see *quality assurance*.

Change management see *business process engineering*.

Chief information officer (CIO) one of the many titles of the individual who leads an enterprise IT organization.

Chief knowledge officer (CKO) businesses where information assets are managed for competitive advantage often employ an executive to oversee the associated knowledge management practices.

Commitment document see *project commitment*.

Critical success factors those performance measures that align most directly with customer satisfaction, i.e., those deliverables most important to the sponsor of the IT service or project.

Customer a direct end user of IT organization products and services.

Customer relationship executive (CRE) the IT liaison responsible for understanding the overall needs of particular customer groups, communicating IT accomplishments in relation to these requirements, and managing customer expectations throughout the delivery process.

Enhancement work performed by an IT team or its external partner providers to extend existing IT services to a new person or population; also, work to extend the functionality or improve the performance of an existing IT product or service.

Enterprise resource planning (ERP) system typically a large, complex suite of software applications that integrate to enable the core business services of the enterprise, e.g., integrating financial, human resources, customer relationship, manufacturing, and distribution management.

Extranet a set of Web services delivered to both external and internal partner providers of an enterprise via the Internet.

Governance a process that involves key stakeholders in the decision-making and performance reviews of the IT organization. These key stakeholders might include internal enterprise customers and their external (paying) customers.

Help desk see *call center*.

Infrastructure the backbone of IT delivery; the networks, communication services, operating systems, servers, desktops, and related platforms, products, and services that provide IT capabilities to the end user.

Internal economy within an enterprise, those frameworks that define resource allocation, distribution, and consumption among business units.

Intranet a Web site designed for the internal use of enterprise personnel for purposes of information sharing and collaboration.

Knowledge management (KM) the process that identifies and brings to bear relevant internal (i.e., from within the enterprise) and external (i.e., from outside the enterprise) information to inform action through collection, translation, cataloging, retrieval, maintenance, and management services.

Knowledge store a content repository in which an organization's documented processes and process outcomes are archived for easy retrieval. As a cornerstone of any knowledge management service, the store serves as the lending library, allowing the organization to leverage content for reuse and redeployment.

Maintenance any activity performed at the discretion of the IT organization that invests in and preserves the value to the customer of an existing IT application or environment, including the following:

- Defect correction — correction of critical defects found in a deployed application that inhibit system availability or performance, e.g., responding to production calls for batch systems running overnight or installing system bug fixes
- Retooling — any change required to a business application due to an upgrade of an infrastructure product, e.g., maintenance necessitated by changes in the production version of PeopleSoft, Lotus Notes, etc.
- Asset protection — business application upgrades in keeping with the vendor releases, e.g., the release of a new version of Microsoft Office or SQL Server
- Disaster recovery procedures — supporting business units in developing their disaster recovery plans and participating in any disaster recovery testing
- Required by external agencies — activities required by external and internal audit, federal agencies, etc.; process to move and maintain applications into a formal change management environment
- Applied research and feasibility analysis — assessments of IT products, services, and processes to determine their appropriateness in line with business unit needs
- Infrastructure and related production support — work associated with the implementation of business system applications and end-user desktops
- System support — Tier 2 and 3 support associated with both the break/fix of existing business application systems and the problems encountered by end-user customers in using these systems
- Information security — identification, assessment, management, and remediation of threats, vulnerabilities, and risks to the electronic information assets of the institution

Master project schedule a tool for reflecting at a given point in time the current status, interdependencies, and resource contentions among IT projects.

Metrics those measures employed as part of service and project delivery to quantify the quality of deliverables in keeping with customer requirements.

Operational readiness see *business process engineering.*

Operational service delivery agreement (OSDA) the standards for the coordination and delivery of work within the IT organization; the formal, internal principles of operation and commitment within the IT team.

Operations report a tool for reflecting the status on a regular basis (e.g., monthly) of IT organization service delivery and project work. This tool may be employed both for the internal management of the IT organization and for communicating with customers outside the IT unit.

Partner provider an IT colleague who is either a downstream provider to, or an upstream user of, IT service or project delivery efforts.

Performance metrics see *metrics*.

Portal see *Web portal*.

Portfolio in this particular context, a grouping of related IT service delivery and project activities, such as the portfolio of work for a particular business unit or a particular Project Management Office staff member.

Power users those individuals inside and outside the IT organization, who due to their knowledge, skills, and business requirements, employ enterprise IT in a sophisticated manner, typically pressing technologies to the limits of their capacity.

Problem resolution (incident) request a customer or partner provider request to support or repair an existing, nonfunctioning IT product or service.

Project commitment a process (and typically a document) that governs the system development life cycle for a particular project, encompassing all new IT asset project work, as well as those few systems or Web site enhancements that are greater than the financial threshold set for enhancement work. Project work will entail the purchase costs of new system or Web site hardware and software, as well as internal and external labor costs, initial product licensing, etc. Once a project deliverable is in production, its ongoing cost is added to the appropriate SLA for the coming year of service delivery.

Project director the IT party responsible for project delivery and the overall coordination of internal and external IT resources.

Project engineering framework (PEF) see *software development life cycle (SDLC)*.

Project management a business process that focuses on the coordination of human and financial resources to ensure the delivery of a product or service in keeping with customer requirements while avoiding or mitigating any risks associated with the effort.

Project manager the staff support person who develops and maintains project commitment documents and plans, facilitates and coordinates project activities, prepares project status reports, manages project meetings, etc.

Project request a service request that, due to its scope, costs, complexity, or associated risks, is to be managed as a separate, formal project.

Prototype a simulated or scaled-down version of an IT deliverable built quickly, at low cost, and with little risk in order to demonstrate the validity of a more substantial IT undertaking. Especially in the world of Web development, prototyping is employed to represent end-user experiences before more time and resources are invested in development.

Pull and push Web services pull services on a Web site are those identified and selected by the user; push services are those delivered by the Web site itself without action by the end user, but usually in line with user profile information.

Quality assurance (QA) a series of processes that test IT applications (systems and services) prior to their release into production. The extended process involves the creation of test scripts based on the business and performance specification of the system, the iterative testing of the product or service, certification that the IT product or service meets or exceeds customer requirements, and management of the release of that deliverable into production and use, including source code management, backup procedures, and so forth.

Reengineering see *business process engineering*.

Release management see *quality assurance*.

Repurposing revising and applying information system components, content, workflows, etc., to an application other than the one for which it was originally intended, i.e., reworking a project plan so that it may be applied to a different but similar IT project.

Return on investment (ROI) a measure of value whereby a financial return is calculated for the investment in a particular undertaking. In real-life situations, some ROI components may be readily quantified in terms of revenue gains, cost reductions, or cost avoidance. However, at times the return on a particular investment is less tangible. For example, an investment that improves customer service may contribute directly to customer retention over time, but its only immediate impact may be to increase operating costs. In the area of IT ROI, quantitative measures need to be balanced with qualitative measures.

Reuse to recycle code, content, or other IT delivery components as these are required for related business purposes, e.g., reusing a sales presentation time and time again when presenting the same product to similar audiences.

Risk management a business process for the identification and mitigation of barriers that obstruct or otherwise jeopardize the delivery of an IT product or service.

Scorecard a simple tool for the representation of project delivery status at a given point in time. A good scorecard tells the entire story about where a project stands in a single page or less.

Service level agreement (SLA) document created through an annual process to address work on existing IT assets, including all nondiscretionary (maintenance and support) IT costs, such as vendor-based software licensing and maintenance fees, and the discretionary costs associated with system enhancements below some threshold amount (e.g., $10,000 per enhancement). Typically, SLA work entails the upgrade costs of system or Web site hardware and software, as well as internal and external labor costs, license renewals, and so forth.

Service request a customer or partner provider request — below a certain risk/cost level — for a new service, for the enhancement of an existing service, or for the extension of an existing service to new or added customers.

Software development life cycle (SDLC) the comprehensive process and framework for managing the delivery of an IT system from its initial framing and scoping through its release into production, including such phases of work as commitment management, requirements analysis, product design, infrastructure readiness preparation, product development, testing and certification, launch and release management, and the sunsetting of superannuated products.

Sponsor the executive leader of the business unit who ultimately approves the funding for the business unit's IT work.

Strategic plan a near- to long-term representation of the steps to be taken by the enterprise to achieve its business objectives, usually a three-to five-year view of goals, objectives, approaches, and success measures or targets.

Taxonomy a uniform, structured system of language applied within a knowledge management process to categorize and catalog content for effective and efficient retrieval by the KM system's user community.

Total cost of ownership (TCO) those elements that, taken together, represent the complete cost associated with an IT acquisition. The TCO includes maintenance and support costs, staff training, business process retooling, and the like, as well as the actual hardware and software costs associated with purchasing an IT product or service.

User see *customer*.

Value chain taken from the model of a manufacturing process, those activities that go into the inception, design, development, delivery, and servicing of a product or service. Like a real link chain, the value chain involves overlapping, connected, and mutually dependent links that culminate in the creation of customer value.

Web asset inventory a tool that maps the various components of a Web site's Web pages, including key technical and usage attributes, to facilitate the development of a Web experience in keeping with customer requirements.

Web portal an Internet site that aggregates content and services for the so-called one-stop-shopping convenience of its users.

Working client those business unit managers who work alongside their IT counterparts to define, develop, and deliver IT products and services to the business unit.

APPENDIX K:
SELECTED READINGS

INFORMATION TECHNOLOGY ORGANIZATIONS — DESIGN AND OPERATIONS

Boar, B.H., *The Art of Strategic Planning for Information Technology*, John Wiley & Sons, New York, 1993.

Davenport, T.H., *Process Innovation: Reengineering Work Through Information Technology*, Harvard Business School Press, Boston, 1993.

Gardner, C., *The Valuation of Information Technology*, John Wiley & Sons, New York, 2000.

Meyer, N.D., *Road Map: How to Understand, Diagnose, and Fix Your Organization*, N. Dean Meyers and Associates, Ridgefield, CT, 1997.

Meyer, N.D., *Structural Cybernetics: An Overview*, N. Dean Meyers and Associates, Ridgefield, CT, 1995.

Mintzber, H., and Quinn, J.B., *The Strategy Process: Concepts and Contexts*. Prentice Hall, Englewood Cliffs, NJ, 1992.

Murphy, T. A portfolio planning approach for IT investment, *Enterprise Operations Management Journal*, 42-10-50 August/September, 2003.

Opper, S., and Fersko-Weiss, H., *Technology for Teams*, Van Nostrand Reinhold, New York, 1992.

Porter, M.E., *Competitive Strategy: Techniques for Analyzing Industries and Competitors*, Free Press, New York, 1980.

Secretariat, Treasury Board of Canada, *Blueprint for Renewing Government Services Using Information Technology*, Treasury Board, Ottawa, 1994.

KNOWLEDGE ORGANIZATIONS, MODELS, DESIGNS, EXAMPLES

American Productivity and Quality Center, *Knowledge Management Consortium Benchmarking Study: Best Practices*, American Productivity and Quality Center, Houston, 1996.

Davenport, T.H., and Prusak, L., *Working Knowledge: How Organizations Manage What They Know*, Oxford University Press, New York, 1997.

Myers, P.S., *Knowledge Management and Organizational Design*, Butterworth-Heinemann, Boston, 1996.

Quinn, J.B., *Intelligent Enterprise*, Free Press, New York, 1992.

Senge, P.M., *The Fifth Discipline: The Art & Practice of the Learning Organization*, Doubleday, New York, 1990.

Zuboff, S., *In the Age of the Smart Machine: The Future of Work and Power*, Basic Books, New York, 1988.

KNOWLEDGE MANAGEMENT PROCESSES

Borgman, C.L., *From Guttenberg to the Global Information Infrastructure: Access to Information in the Networked World*, MIT Press, Cambridge, MA, 2000.

Chaffey, D., *Groupware, Workflow and Intranets: Reengineering the Enterprise with Collaborative Software*, Digital Press, Woburn, MA, 1998.

Donovan, J.J., *Business Reengineering with Technology: An Implementation Guide*, Cambridge Technology Group, Cambridge, MA, 1993.

Fruin, W.M., *Knowledge Works: Managing Intellectual Capital at Toshiba*, Oxford University Press, New York, 1997.
Honeycutt, J. *Knowledge Management Strategies*, Microsoft Press, Redmond, WA, 2000.
Ranadive, V., *The Power of Now*, McGraw-Hill, New York, 1999.
Skikantaiah, T., and Koenig, M.E.D., Eds., *Knowledge Management*, ASIS, Medford, NJ, 2000.
Stefik, M., *The Internet Edge: Social, Technical and Legal Challenges for a Networked World*, MIT Press, Cambridge, MA, 1999.

KNOWLEDGE MANAGEMENT PLATFORMS

Berson, A., Smith, S., and Thearling, K., *Building Data Mining Applications for CRM*, McGraw-Hill, New York, 2000.
Blum, D., *Understanding Active Directory Services*, Microsoft Press, Redmond, WA, 1999.
Covill, R.J., *Implementing Extranets: The Internet as a Virtual Private Network*, Digital Press, Woburn, MA, 1998.
Freedman, A. *The Internet Glossary and Quick Reference Guide*, AMACOM, New York, 1998.
Inmon, W.H., and Caplan, J., *Information Systems Architecture*, QED Publishing Group, Wellesley, MA, 1992.
Kimball, R. and Merz R., *Data Webhouse Tool Kit: Building the Web-Enabled Data Warehouse*, John Wiley & Sons, New York, 2000.
Mattison, R., *Web Warehousing and Knowledge Management*, McGraw-Hill, New York, 1999.
Mena, J., *Data Mining Your Website*, Digital Press, Woburn, MA, 1999.
Muller, R.J., *Database Design for Smarties: Using UXL for Data Modeling*, Morgan Kaufmann Publishers, San Francisco, 1999.
Quinn, J.B., Zien, K.A., and Baruch, J.J., *Innovation Explosion: Using Intellect and Software to Revolutionize Growth Strategies*, Free Press, New York, 1997.
Ruggles, R.L., Ed., *Knowledge Management Tools*, Butterworth-Heinemann, Boston, 1997.
Schmeiser, L., *The Complete Web Site Upgrade and Maintenance Guide*, Sybex, Alameda, CA, 1999.
Sonnenreich, W. and Macinta, T., *Web Developer.Com Guide to Search Engines*, John Wiley & Sons, New York, 1998.
Tannenbaum, A., *Implementing a Corporate Repository*, John Wiley & Sons, New York, 1994.
Trepper, C., *E-Commerce Strategies: Mapping Your Organization's Success in Today's Competitive Marketplace*, Microsoft Press, Redmond, WA, 2000.
White, T.E. and Fischer, L., *The Workflow Paradigm*, Future Strategies, Alameda, CA, 1994.

PROJECT MANAGEMENT

Black, R., *The Complete Idiot's Guide to Project Management with Microsoft Project 2000*, QUE Press, New York, 2000.

Chang, Y.S., Labovitz, G., and Rosansky, V., *Making Quality Work*, HarperCollins, New York, 1993.
Friedlein, A., *Web Project Management*, Morgan Kaufmann Publishers, San Francisco, 2001.
Kerzner, H., *Strategic Planning for Project Management Using a Project Management Maturity Model*, John Wiley & Sons, New York, 2001.
Kesner, R.M., *Information Systems: A Strategic Approach to Planning and Implementation*, American Library Association, Chicago, 1988.
Lewis, J.P., *The Project Manager's Desk Reference*, 2nd ed., McGraw-Hill, New York, 1999.
Murch, R., *Project Management Best Practices for IT Professionals*, Prentice Hall, Upper Saddle River, NJ, 2000.
Philips, J., *IT Project Management: On Track from Start to Finish*, McGraw-Hill/Osborne Media, New York, 2002.
Project Management Institute, *A Guide to the Project Management Body of Knowledge, 2000 Edition*, Project Management Institute, Newtown Square, PA, 2000.
Verma, V.K., *Organizing Projects for Success*, Project Management Institute, Upper Darby, PA, 1995.

PROJECT MANAGEMENT OFFICE

Debuzman, M., *The Project Management Office: Gaining the Competitive Edge*, ESI International, Arlington, VA, 1999.
Englund, R.L., *Creating the Project Office: A Manager's Guide to Leading Organizational Change*, Jossey-Bass, San Francisco, 2003.
Fisher, K., *Leading Self-Directed Work Teams*, McGraw-Hill, New York, 1993.
Frame, J.D., et al., *The Project Office: Best Management Practices*, Crisp Publications, Normal, IL, 1998.
Hallow, J.E., *The Project Management Office Tool Kit*, AMACOM, New York, 2001.
Miranda, E., *Running the Successful Hi-Tech PO*, Artech House, Norwood, MA, 2003.
Rad, P.F., Levine, G., and Kiniry, J.R., *The Advanced Project Management Office: A Comprehensive Look at Function and Implementation*, CRC Press, Boca Raton, FL, 2002.
Verma, V.K., *Managing the Project Team*, Project Management Institute, Newtown Square, PA, 1997.

SERVICE DELIVERY MANAGEMENT

Assirati, B., Ed., *Service Delivery (IT Infrastructure Library Series)*, Her Majesty's Stationery Office, London, 2001.
DiPasquale, T., et al., *IT Services: Cost, Metrics, Benchmarking and Marketing*, Prentice Hall, Upper Saddle River, NJ, 2001.
Melik, R., et al., *PSA: Professional Services Automation: Optimizing Project and Service Oriented Organizations*, John Wiley & Sons, New York, 2002.
Sturm, R., Morris, W., and Jander, M., *Foundations of Service Level Management*, SAMS, Indianapolis, IN, 2000.
Walker, G. and Walker, G.S., *IT Problem Management*, Prentice Hall, Upper Saddle River, NJ, 2001.

THE SOFTWARE DEVELOPMENT LIFE CYCLE (SDLC)

Bass, L., Clements, P., and Kazman, R., *Software Architecture in Practice*, Addison-Wesley, Reading, MA, 1999.

Beck, K., *Extreme Programming Explained*, Addison-Wesley, Reading, MA, 2000.

Boar, B.H., *Constructing Blueprints for Enterprise IT Architectures*, John Wiley & Sons, New York, 1999.

Cantor, M.R., *Object-Oriented Project Management with UML*, John Wiley & Sons, New York, 1999.

Fowler, M., *Refactoring: Improving the Design of Existing Code*, Addison-Wesley, Reading, MA, 1999.

Hohmann, L., *Beyond Software Architecture: Creating and Sustaining Winning Solutions*, Addison-Wesley, Reading, MA, 2003.

McConnell, S., *Code Complete*, Microsoft Press, Redmond, WA, 1993.

McConnell, S., *Software Project Survival Guide*, Microsoft Press, Redmond, WA, 1997.

INDEX

A

Action (annual) plan, 73–88, 247–262, 317–320
  Action (annual) plan off-cycle process, 86–87
  Action (annual) plan process, 82–86, 231–233
Artifacts (project), 183–187, 247–252, 271–282, 301–304, 305–316
Architecture (IT), management of, 199–202, 214–220
  Architecture planning process, 203–207
  Architecture planning team, 207–209
Alignment, 57–60, 253–262, 271–282, 295–300

B

Business analyst, 40–41, 147–170
Business environments, *see* Enterprise
Business priorities, setting of, *see* Prioritization
Business process engineering, 132–135, 150–167
Business process mapping, 154–158, 178–184
  Business process mapping, roles and responsibilities matrix, 158–163
  Business process mapping, performance metrics, 164–165
  Business process mapping, process rules, 163–164
Business requirements gathering, 150–154, 166–170, 184–186, 203–207, 247–252, 283–294

C

Call center, 93–99, 103–107, 233–235
Certification, *see* Quality assurance.
Change management, *see* Business process engineering.
Chief information officer (CIO), 231–233, 253–262, 295–301
Commitment document, *see* Project commitment process
Critical success factors, xxvi–xvii, 11, 21, 77–78, 80–81, 295–300, 301–304
Customer relationship executive (CRE), 11, 14–15, 17, 35–37, 93–94

E

End user, *see* Sponsor, and Working Client
Enhancements, serving of, 102–103, 105–108, 271–282
Enterprise, xxi–xxvi, 147–150
Enterprise, resource planning (ERP) system, xxiii–xxiv
Enterprise, transformation diamond, xxiv–xxvi, 61–63, 65,

G

Governance, 57–72, 253–262, 295–300, 305–316

H

Help desk, *see* Call center

335

I

Infrastructure, 168–170, 203–207, 214–219
Information technology (IT) organization, xv–xvi, 1–3, 26–28, 30–33, 171–178
  Information technology organization, product and service inventories, 63–65, 105–107
  Information technology organizations, roles and responsibilities, 33–34, 239–241
  Information technology organizations, weaknesses of, 28–30, 174–178, 203–206
Internal economy, 3–7, 58–59

K

Knowledge management, 49–50, 151–154, 171–198, 199–201, 238–239
Knowledge store, 182–184, 186–193, 203–220

M

Maintenance, servicing of, 101–103, 105–107, 271–282
Master project schedule, 73–88, 317–320
Metrics, 20–24, 50–56, 95–100, 108–112, 164–165, 214–219
Murphy's portfolio scoring tool, *see* Portfolio management

N

New England Financial (now part of MetLife), 202–203
Northeastern University, 174–178

O

Operational service delivery agreement (OSDA), 103–112; *see also* Service level agreement (SLA)
Operations report, 23, 142–144
Osborne/Kesner "enterprise" transformation diamond, *see* Enterprise, transformation diamond

P

Performance metrics, *see* Metrics
Planning, *see* Action (annual) plan
Portal, *see* Web portal.
Porter's value chain model, *see* Value chain
Portfolio management, 63–65, 118–120, 139–143, 214–219, 317–320
Power users, *see* Sponsor, and Working Client
Prioritization, 60–88, 247–252, 301–304, 317–320
Problem resolution (incident) request, 107–113, 271–282
Project commitment process, 6, 123–124, 129–139, 283–294, 305–316
Project delivery tools, ix–xiv, xvi–xvii, 247–252, 283–294, 301–316
Project director, 18, 38–39, 130–131, 139–143
Project delivery, 5–6, 8–10, 16–20, 115–146, 236–238, 283–294, 301–304, 305–320
Project engineering framework (PEF), *see* Software development lifecycle (SDLC)
Project manager, 18, 39–40, 139–146
Project management, 19–20, 47–48, 118–129, 131–139
  Project managing staffing, 137–139
  Project management reporting, 139–144
Project management office (PMO), 30–33, 42–50,
  Project management office, roles and responsibilities, 44–47, 51–55, 113–114, 144–146, 194–195
  Project management office, value proposition, 50–55, 196–198, 220–223, 225–243, 263–270
Project request, 67–72, 115–128, 247–252, 305–316
Prototype, 125, 184–195, 207–213
Pull and push web services, 187–188, 207–213; *see also* Web services

Q

Quality assurance, 108–112, 124, 139–143, 193–194, 218–220

R

Reengineering, *see* Business process engineering
Release management; *see also* Quality assurance
Reporting, 22–24, 108–112, 139–143, 167–168, 196–198, 214–219; *see also* Metrics
Return on investment (ROI), 2–3, 18, 51–55, 66–67, 220–223, 225–243, 263–270
Risk management, 135–137, 301–304

S

Scorecard, 139–142
Service delivery, 4–5, 7–9, 10–16, 89–93, 233–236, 271–282
Service delivery manager, 37–38
Service delivery tiers, 98–100
Service delivery tools, ix–xiv, xvi–xvii, 100–103, 271–282
Service level agreement (SLA), 5–6, 13–14, 17, 93–114

Service request, 89–93, 98–100, 247–252, 271–282,
Software development lifecycle (SDLC), 16, 116–117, 120–123, 184–192, 217
Sponsor, 34, 103–105, 129–134, 158–162
Strategic plan, 73–88, 253–262, 317–320

T

Taxonomies, 183–184, 203–207
Testing, *see* Quality assurance
Total cost of ownership (TCO), 58, 66–67, 72–73, 87–88, 225–237, 247–252

U

Users, *see Sponsor*, and Working Client

V

Value chain, xxi–xxiv

W

Web asset inventory, 182–183, 207–213
Web portal, 178–179, 186–193, 214–220
Web services, 186–193, 209–214
Working client, 34–35, 103–105, 129–134, 158–162